"AI: Work Smarter and Live Better Within Civilization's Next Big Disruption *is a valuable guide for anyone who wants to understand and leverage AI in their personal and professional lives. With insights from 33 global experts, this book shows AI's impact on industries and society, empowering readers to stay ahead of the curve. If you want to thrive in an AI-driven world, start here.*"

—JARVIS LEVERSON,
Founder of Accelerated Growth Club

"This book is not just about understanding AI—it's about preparing yourself for a future where AI is deeply integrated into every aspect of life. The authors do a brilliant job of presenting diverse insights from professionals across the globe, making complex AI concepts accessible to everyone. If you want to work smarter and live better in the age of AI, you need to read this book."

—JIM ROBERTSON,
Creator of *Working in the Age of AI*

"This book, AI, offers a practical and clear-eyed look at how AI is transforming industries and reshaping human life. With expert contributions from various fields, this book provides actionable insights that will help readers adapt and succeed as AI becomes more pervasive. Highly recommended for anyone wanting to stay ahead in the AI era."

—NANCY CARTER,
Postdoctoral AI Researcher

AI

AI

Work Smarter and Live Better Within Civilization's Next Big Disruption

Authored by:

Erik Seversen, Dimitrios A. Alexandrou, PhD, Robert Bahr, Katerina Bourdoukou, Michael W. Bradicich, Sam Calhoun, Matt Collette, Dave Cook, Jean-Michel JCA Davault, Malcom W. Devoe, PhD, Hassen Dhrif, PhD, Parham Emami, Craig Fulton, Jeremy Kofsky, Brandon Lester, Peter Lundgreen, Veejay Madhavan, Ghofran Massaoudi, Grzegorz P. Mika, Salah Aldin Mokhayesh, Louis Mono, Greg Ombach, PhD, Ivan Padabed, Tanmay Patani, Greg Pellegrino, Keith Pham, PhD, Nikki Remkes, Gabriel Lars Sabadin, PhD, Cedric De Schaut, Stanislav Sorokin, Matthew Switzer, Sakina Syed, Susan O. Tejuosho, and Martin Wyss, PhD

THIN LEAF PRESS | LOS ANGELES

AI: Work Smarter and Live Better Within Civilization's Next Big Disruption. individual chapters. Copyright © 2025 by Dimitrios A. Alexandrou, PhD, Robert Bahr, Katerina Bourdoukou, Michael W. Bradicich, Sam Calhoun, Matt Collette, Dave Cook, Jean-Michel JCA Davault, Malcom W. Devoe, PhD, Hassen Dhrif, PhD, Parham Emami, Craig Fulton, Jeremy Kofsky, Brandon Lester, Peter Lundgreen, Veejay Madhavan, Ghofran Massaoudi, Grzegorz P. Mika, Salah Aldin Mokhayesh, Louis Mono, Greg Ombach, PhD, Ivan Padabed, Tanmay Patani, Greg Pellegrino, Keith Pham, PhD, Nikki Remkes, Gabriel Lars Sabadin, PhD, Cedric De Schaut, Stanislav Sorokin, Matthew Switzer, Sakina Syed, Susan O. Tejuosho, and Martin Wyss, PhD

Disclaimer—The advice, guidelines, and all suggested material in this book is given in the spirit of information with no claims to any particular guaranteed outcomes. This book does not replace professional consultation. Anyone deciding to add physical or mental exercises to their life should reach out to a licensed medical doctor, therapist, or consultant before following any of the advice in this book; anyone making any financial, business, or lifestyle decisions should consult a licensed professional before following any of the advice in this book. The authors, publisher, editors, and organizers do not assume and hereby disclaim any liability to any party for any loss, damage, or disruption caused by anything written in this book.

Library of Congress Cataloging-in-Publication Data
Names: Seversen, Erik, Author, et al.
Title: *AI: Work Smarter and Live Better Within Civilization's Next Big Disruption.*
LCCN: 2025902870

ISBN 978-1-953183-73-6 (hardcover) I 978-1-953183-72-9 (paperback)
ISBN 978-1-953183-71-2 (eBook) I 978-1-953183-74-3 (audiobook)

Artificial Intelligence, Science and Technology, Business, Professional Development
Cover Design: 100 Covers
Interior Design: 100 Covers
Editor: Nancy Pile
Thin Leaf Press
Los Angeles

THIN
LEAF

Thank you for reading this book. There is information found within the following pages that can greatly benefit your life, but don't stop there. Make sure you get the most you can from this book and reach out directly to the expert-authors who want to help you reach your goals in an AI-driven world, to thrive within civilization's next big disruption, and to manifest success in your life. Contact information for each author is found at the end of their respective chapter.

To the great minds who are figuring out how to create super-intelligent machines and to those who are aiming to protect civilization from them.

CONTENTS

INTRODUCTION

By Erik Seversen
Author of *Ordinary to Extraordinary*
Los Angeles, California

The other day, I was walking in a neighborhood near my home in Los Angeles. As I crossed the street, I saw a Waymo driverless car approaching me. With neither driver nor passenger in the car, I had to make a choice. *Do I trust this car to stop without hitting me?* While I've seen driverless cars multiple times (a daily occurrence now), I could see the car's cameras and spinning thing on top, and I believe autonomous cars are probably better drivers than humans—I still felt a strange sensation in my stomach that I had two choices: walk super-fast to get out of the car's path or trust my life to a machine that knows the rules of the road, but that really doesn't care if I live or die.

I didn't speed up, but I watched cautiously, my human eyes following the machine, trying to understand what the vehicle was thinking or what input the car had learned from to access and react to the immediate situation at hand, i.e., an "object" (me) deemed to be important based on the large language model it learned from. In a weird sense, I actually felt empathy for the car when I realized that while it *knew* what to do to avoid the "object" in its path, it *felt* nothing.

A few days later, I was using ChatGPT. I prompted Alvin (the name which I've given my ChatGPT) to create an outline for a series of blog posts for me. The program spit out ten points within seconds and asked if I would like any of the points expanded upon. I wrote back, "Yes, please list xyz information about point seven, and I'll work on the others later." Alvin spit out an expanded point seven and then added, "It'll be fun developing the other points with you later."

Wow, I thought. *The computer program I'm using just used the word "fun," an emotion word, to describe something.* Again, I had the strange sensation that other than learning from inputs given by the ChatGPT engineers, this program doesn't have any emotions at all, but it made me feel good because of the words it used while interacting with me. So, a non-sentient being was able to affect my feelings. I had to pause to reflect on this, and then I pushed the thoughts out of my head, so I could get back to work.

While I gave those two anecdotes showing my human interaction with AI, this is just scratching the surface. AI isn't coming, AI is here ... and it's just getting started. I strongly believe that within a short number of years, AI is going to be present in almost every aspect of our lives. Certainly, anything dealing with electronics or computing will be AI-driven. Humans are going to need to figure out their association with AI as civilization's next big disruption.

I'm not an AI expert, but having had hundreds of conversations with AI professionals in multiple fields of study and enterprise, I know enough about AI now to be very excited about what is happening with new technologies including autonomous AI agents, neuro radiance fields, and quantum machine learning, to name a few. I also know enough to be scared of what is happening (and could happen) with AI as machines approach the singular point where they are able to teach themselves. As humans embark on what many are calling the "next stage of human social evolution," I believe things will change more rapidly than many people will be able to keep up.

I'm not here to scream about AI stealing people's jobs or taking over the world, but I am here to let you know that if you don't pay attention to what is happening with AI, you will be at a disadvantage. However, there is also some good news. If you are open to seeing possibilities of AI saving time and energy, you might be among those who are able to work smarter and live better as humans and machines begin to work more closely together. The goal of this book is to help you realize what is happening with AI, so you can choose whether you want to use AI to your advantage as the AI revolution changes the world.

With this goal in mind, I want to thank the authors of this book who agreed to share their knowledge of AI in a vast array of industries. These individuals have real-world experience creating, using, and governing AI. They are at the forefront of the AI discussion, and they are willing to share their knowledge and AI experience with you.

While many people think AI is basically Copilot, ChatGPT, and Gemini, there is a lot more to it. The goal of the first AI book I produced, *The AI Mindset*, was to help people create a mindset around what is happening with AI, a mindset that is excited and optimistic about the many great advances in AI, advances that just might allow humans the freedom to live more balanced lives. However, the AI mindset is also a mindset of caution as we need to ensure that AI is fair, non-biased, and advances good rather than harm. Furthermore, I think it would be naïve not to think about the possibility of AI actually taking over or destroying humanity. While this sounds like science fiction, all of the smartest people (from Stephen Hawking, Stuart Russell, Elon Musk, Bill Gates, to Sam Altman) discuss this existential threat as very real.

While touched upon, thoughts about machine sentience and AI as an existential threat are beyond the scope of this book. The goal of this book, the second in the AI series I produced, is simply to help introduce the many ways AI is becoming integrated in various industries in the world, so people can make educated decisions about how they want to interact with AI in their life.

Since I'm not an AI expert, I didn't try to write this book by myself. Rather, I reached out to a number of people from around the world who know much more about AI than I do. While curating this book, I've had the pleasure of learning about AI in government, education, law, economics, finance, healthcare, cybersecurity, engineering, digital transformation, and more. I've learned about blockchain technology, cloud services, large language models, generative AI, machine learning, automation, chatbots, and data science. Yes, these are just some of the concepts found in the chapters that follow. With this book, anyone aiming to work smarter and live better within civilization's next big disruption will better understand AI and will gain a vast array of ideas that will help them become more aware of what AI has to offer and how it is being used.

In order to create the best book possible, I solicited the help of 33 AI experts from various backgrounds and locations. The co-authors of this book come from all over the USA, Canada, Sweden, Finland, Denmark, the Netherlands, Belgium, France, Germany, Switzerland, Italy, Poland, Portugal, Tunisia, Singapore, Israel, and Greece.

These authors are professionals who are university professors, business owners, consultants, product managers, engineers, software developers, cybersecurity leaders, advisors, TEDx and keynote speakers, researchers, platform architects, senior data analysts, IT directors, CTOs, and more. The one thing these individuals have in common is that they all have something to share about artificial intelligence, and these ideas are available to you now.

Although this book is organized around the united theme of working smarter and living better within civilization's next big disruption, each of the chapters is totally stand-alone. The chapters in the book can be read in any order. If you really want to get a survey of what is going on with AI, the glossary at the end of the book might be a good introduction to some of the nuts and bolts of AI. I encourage you to look through the table of contents and begin wherever you want. However, I urge you to read all the chapters because, as a whole, they provide a great array of perspectives. Each is valuable in helping you understand how AI is being used in all areas of human industry.

It is my hope that you discover something in this book that helps you navigate civilization's next big disruption as humans and machines become more interdependent and AI technology continues to insert itself into many aspects of our everyday lives.

About the Author

Erik Seversen is on a mission to inspire people. He holds a master's degree in anthropology and is a certified practitioner of neuro-linguistic programming. Erik draws from his years of teaching at the university level and years of real-life experience in business to motivate people to take action, creating extreme success in business and in life.

Erik is a TEDx and keynote speaker who has reached over one million people through his public speaking and live courses. He has visited 99 countries and all 50 states in the USA and has climbed the highest mountains on four continents, 15 countries, and 18 states. Erik has published 17 bestselling books on the topics of mindset, success, and peak performance, and he has helped over 400 people become best-selling authors. He is a full-time writer, book consultant, and speaker, and he lives by the idea that success is available to everyone—that living an extraordinary life is a choice.

Erik lives in Los Angeles with his wife and has two boys currently studying at university.

Contact Erik for interviews, speaking, or book publishing consultation.
Email: Erik@ErikSeversen.com
Website: www.ErikSeversen.com
LinkedIn: https://www.linkedin.com/in/erikseversen/

CHAPTER 1

EMPOWERING GOVERNANCE WITH AI: CITIZEN-CENTRIC SOLUTIONS IN THE GOVTECH REVOLUTION

By Dimitrios A. Alexandrou, PhD
Co-Founder & Business Innovation Director
Athens, Greece

> *Big results require big ambitions.*
> —Heraclitus

In the rapidly evolving landscape of the digital age, artificial intelligence (AI) emerges as both the loom and the thread, weaving unprecedented possibilities into the essence of human governance. AI has become a transformative force, reshaping the dynamics of how governments engage with and provide services to their citizens. At its core, AI exceeds the boundaries of traditional bureaucratic systems, offering to governments and official authorities instruments to become agile, responsive, and

1

profoundly citizen-centric. AI acts as a bridge between the complexities of public administration and the simplicity citizens seek, harnessing the power of data, computational power, and natural language processing to revolutionize public services.

We have to imagine a government not as a monolithic entity, burdened by inefficiencies and bureaucracy, but as a living, breathing organization—one that anticipates the needs of its citizens, addresses their queries in real time, and delivers tailored solutions with a level of precision that was once unimaginable. From automated service delivery software platforms that facilitate the execution of intricate bureaucratic processes to chatbots that converse as seamlessly as humans, AI transforms entirely the paradigm of citizen interaction with the authorities. It enables a paradigm shift from reactive to proactive administration, paving the way for governance that is not just accessible, but anticipatory.

This specific chapter delves into the remarkable strides performed by the government of the Hellenic Republic, where AI has been utilized as a beacon of progress in serving its citizens. By exploring real-world applications like mAIgov, mAiGreece, and CadastreBot, we unveil how AI uplifts governance into an art—an art that empowers, engages, and redefines the relationship between citizens and their state.

Embarking to the AI Era

Fifteen months ago, the government of the Hellenic Republic embarked on an ambitious voyage into the boundless horizon of artificial intelligence. Having a vision rooted in innovation and a steadfast commitment towards its citizens, this journey marked the dawn of a new era in governance. The government undertook a meticulous exploration of the AI landscape, examining cutting-edge technologies and identifying opportunities to transform public services. What began as an aspiration quickly evolved into a focused endeavor to redefine the essence of "citizen-state interaction."

It was not a transformative route without challenges. Guided by a spirit of research and a relentless pursuit of excellence, the government of the Hellenic Republic embarked on the complicated and intriguing process of designing and implementing AI-driven solutions. The result? A flourishing ecosystem of AI initiatives that prioritize accessibility, efficiency, and empowerment—ushering Greece into the AI era with purpose and resolve.

The Technical and Operational Reefs

LLM Model Resilience to Prompt Attacks

One significant challenge the government of the Hellenic Republic encountered during the implementation of AI chatbots was ensuring the
resilience of the large language models (LLMs) to "prompt attacks."
These specific attacks take advantage of the flexibility of natural language
interfaces to manipulate model behavior, often generating unintended
or malicious outputs. In the context of governmental services, where
accuracy and trust are paramount, such vulnerabilities could significantly compromise the reliability of information provided to citizens
and professionals.

Prompt attacks may occur in various forms, such as adversarial
inputs designed to bypass safeguards, misleading questions that prompt
incorrect responses, or injection of malicious commands intended to
manipulate downstream behavior of the LLM. For example, in mAIgov,
citizens rely on accurate information about governmental services, so an
error induced by a prompt attack could spread misinformation. Ensuring
the resilience of the LLMs required implementing robust countermea-
sures to prevent misuse.

Heterogeneous and Unstructured Knowledge Sources

One of the most significant challenges encountered during the development of the Hellenic government's AI chatbot initiatives was the acquisition and processing of heterogeneous and unstructured knowledge
sources. The information required to power the chatbots—mAIgov, mAi-
Greece, and CadastreBot—is stored in different formats and repositories.
These sources ranged from semi-structured files such as XML and JSON
documents, to entirely unstructured formats, including PDFs, electronic
images, and textual reports.

Unstructured data posed the greater challenge. For example, mAI-
gov had to utilize documents detailing the complete set of governmental
business processes often stored as lengthy, text-heavy PDFs, as well
as electronic services descriptions, which were much less rich in text.
These documents lacked standardization in format and length, requiring
homogenization and curation based on automated summarization so as
to produce embeddings of similar length that would comprise the knowledge sources of the brain of the chatbot.

Nationwide Unhindered Access to AI Resources
Another significant challenge faced during the development of the Hellenic government's AI chatbots was the enablement of nationwide, unhindered access to the AI resources, given the immense scale of potential users that should be served. With over 10 million citizens and millions of international visitors expected to interact with mAIgov, mAiGreece, and CadastreBot, the challenge lay not only in deploying AI models capable of providing accurate and timely responses but also in ensuring system reliability under potentially unprecedented demand.

The demand for high availability and scalability led to the integration of a robust queuing system to manage simultaneous requests efficiently, taking under consideration the limited capacity of the AI resource. This was critical to prevent system overloads that could lead to delays or downtime, particularly during peak usage times. The queuing system acts as a mediator, managing incoming requests and prioritizing them based on system capacity while ensuring that no user experiences prolonged waiting times.

The Biggest Challenge ... What Else? The Greek Language!
Among all the challenges faced during the implementation of these AI-enabled chatbots, one stood out as both unique and demanding: handling the Greek language! Renowned for its complexity and linguistic richness, Greek presented significant issues that affected the implementation process.

One of the foremost challenges was the cost of handling the Greek language in AI systems. Greek is considered the most "expensive" language in terms of tokens—a crucial factor in natural language processing. A single sentence in Greek can produce significantly more tokens compared to other languages like English. The specific characteristic inflates processing costs and increases the computational load, making efficiency a constant priority. Optimizing token usage without compromising the quality of responses required meticulous fine-tuning of the chatbots' language model.

Another considerable challenge was the process of vectorization—the transformation of words into numerical representations that AI models are able to comprehend. Greek, with its inflectional morphology, rich vocabulary, and unique syntax, is notoriously difficult to vectorize accurately. Words in Greek change form depending on their grammatical role in a sentence, creating a vast array of variations for even simple concepts. Capturing the subtleties of these transformations and ensuring that the embeddings generated by the AI were contextually meaningful demanded a sophisticated approach and large training dataset.

The searching process in Greek further compounded the difficulty. Discovering appropriate embeddings that matched user queries with relevant responses often faltered due to nuances in phrasing and spelling. Unlike English, where queries and results tend to align straightforwardly, Greek posed challenges in recognizing synonyms, idiomatic expressions, and variations of the same root word. These issues required the implementation of advanced semantic search techniques tailored specifically to the intricacies of the Greek language.

Taming the Waves

The voyage into the AI era was anything but smooth sailing. As the government of the Hellenic Republic entered into uncharted waters, it encountered a set of challenges that tested both resolve and ingenuity. And fortunately: everything that could go wrong. did!

Strict Prompting Towards Closed-World Assumption

In our UBITECH engineering team's pursuit of precision, quality, and reliability within the AI initiatives of the government of the Hellenic Republic, we adopted a robust closed-world assumption approach. At the heart of the specific methodology lies the principle that AI models must operate exclusively within the boundaries of predefined and authoritative knowledge, ensuring every response remains firmly tethered to the truth.

This approach is underpinned by retrieval-augmented generation (RAG), a sophisticated framework that ensures model outputs are informed solely by official knowledge repositories provided by the government of the Hellenic Republic. By incorporating a retrieval layer, the system accesses and utilizes curated datasets, effectively mitigating the risk of "hallucinations"—the generation of unsupported or erroneous information. This fact guarantees that the responses remain authoritative and relevant, safeguarding both accuracy and credibility in citizen interactions with the state.

—

Complementing this framework, our team enforces strict prompting techniques to solidify adherence to predefined operational guidelines. These prompts establish an immutable set of "do" and "do not" rules, embedding layers of model security that cannot be bypassed by user input. Thus, we ensure that the AI model consistently delivers responses aligned with governmental policies, ethical standards, and the scope of the intended applications.

Knowledge Sources Curation and Summarization
To ensure the effective curation and summarization of knowledge sources, we developed a meticulously designed, multi-layered approach that transforms raw, unstructured data into actionable embeddings. This process begins with optical character recognition (OCR), specifically tailored for table extraction, enabling the identification and isolation of structured data embedded within scanned documents or images. By converting these static elements into dynamic datasets, we establish the foundation for precise and efficient analysis.

Next, we employ semantic and text-length splitting, a critical step in segmenting content into meaningful and manageable units. This ensured that the data remains contextually coherent while avoiding information overload during subsequent processing. Image extraction is then applied to identify and retrieve visual elements—such as charts, graphs, and illustrations—that complement the textual content, enhancing the comprehension of the material.

To refine the curated data, we execute noise removal as well as filtering out superfluous characters, symbols, and irrelevant artifacts that may compromise the quality of analysis. This data curation workflow not only amplifies the quality and usability of extracted content but also paves the way for more accurate and insightful summarization.

Utilization of Software Components to Handle Knowledge Acquisition in Various Languages
The Hellenic government's engagement with AI requires a sophisticated approach to managing knowledge acquisition across multiple languages. Recognizing the inherent challenges of multilingual sources, we implement a hybrid methodology that combines precision with adaptability. The specific approach reassures that no nuance is lost, regardless of the linguistic diversity of the input sources.

At the core of our system lies a dual-processing framework. Text is first translated into English—a lingua franca for consistent and stream-lined processing. Here, English-specific embedding models analyze the translated text, extracting its semantic essence with unparalleled precision. Simultaneously, the original text is processed using advanced multilingual embedding models, designed to capture the unique linguistic and cultural intricacies of each language. This parallel analysis allows us to synthesize insights that reflect both universal meaning and language specificities. By combining the strengths of language-agnostic embeddings with English-specific processing, we achieve a holistic understanding of the data. The hybrid approach not only enhances the system's semantic comprehension, but also ensures a robust handling of diverse linguistic structures. The result is a resilient, multilingual knowledge acquisition framework, capable of handling the complexities of global communication while upholding the highest standards of accuracy.

The Evolution

Our partnership with the government of the Hellenic Republic has been a transformative journey, marked by innovation and a steadfast commitment to harnessing AI for the alleviation of public service delivery. It all began with the launch of mAIgov, in December 2023, an infoBot designed to empower citizens by providing accurate and accessible information about government services in a personalized fashion. This foundational milestone set the stage for a bold evolution.

Two months later, in February 2024, mAIgov actionBot emerged, introducing a groundbreaking capability: the execution of third-party services through the function chaining technique. This advancement transitioned mAIgov from a passive source of information to an active participant in citizen interactions, streamlining tasks and reducing bureaucracy.

In April 2024, we delivered CadastreBot, a specialized solution tailored to the unique demands of the Hellenic Land Registry. By incorporating five distinct personas—citizen, lawyer, notary, engineer, and bailiff—it provides tailored information and support, showcasing the power of personalization in AI-driven governance.

By June 2024, mAiGreece was operational, offering personalized, location-based insights to visitors across the Greek territory. This innovative chatbot enriches the travel experience while highlighting Greece's commitment to tourism excellence.

Finally, in November 2024, mAIgov agentBot took center stage. With the ability to execute complex workflows, it evolves into an integrated service provider, orchestrating intricate electronic processes and delivering seamless citizen experience. This progression symbolizes the Hellenic government's unwavering dedication to pioneering AI solutions that redefine public service delivery.

What More to Expect?

Tomorrow ... is now! As we stand on the edge of a new technological frontier, the evolution of AI is leading us towards multi-agent AI systems—a dynamic orchestration of intelligent agents capable of collaborating in a cross-organizational fashion. The future holds the promise of cross-organizational workflow execution, where disparate governmental bodies seamlessly interconnect to deliver integrated end-to-end services. Imagine a Greek citizen navigating complex bureaucratic processes with the ease of a single interaction—an AI-driven ecosystem that anticipates needs, resolves issues, and unifies fragmented systems into a cohesive whole.

This vision reflects our commitment not just to innovation, but towards a citizen-centric approach that redefines the role of technology in governance. With each milestone, we continue to push boundaries, setting a precedent for how AI can empower governments to provide truly seamless, efficient, and transformative public services. The next destination in this journey is one of limitless possibilities, where the synergy of AI and human ingenuity paves the way for a smarter, more connected future for Greece and beyond.

About the Author

Dimitrios Alexandrou, PhD Eng., is a Greek entrepreneur. He is one of the co-founders and business innovation director of UBITECH responsible for commercial corporate activities, EU and Greek public sector and overseas activities, private sector projects, and IT consulting services. He also leads the business services of AiLabs, a spinoff initiative covering artificial intelligence business initiatives and products.

He received his diploma in electrical and computer engineering from the National Technical University of Athens and holds a PhD from the same school of NTUA in the field of expert systems. With a PhD in self-adaptive healthcare business processes (clinical pathways) his expertise spans business process modeling, expert and rule-based systems, generative artificial intelligence, and large language models. He has been leading technical teams and consortia in the frame of EU co-funded projects, as well as coordinating UBITECH commercial activities.

Email: dalexandrou@ubitech.eu
LinkedIn: https://www.linkedin.com/in/dimitris-alexandrou-ab828a1/

CHAPTER 2

AI IN EDUCATION: EMPOWERING TEACHERS AND TRANSFORMING LEARNING

By Robert Bahr, MA
Educator, AI and Education Technology Consultant
Kirkkonummi, Finland

The dawn of the AI revolution is transforming both teaching and learning. For the first time in history, we can expand our ideas and receive assistance from intelligent machines, not just other people. Computers are no longer passive tools; they actively help us expand our knowledge. However, as machines become smarter, the potential risk of large-scale societal disruption grows as well. This chapter explores the possibilities and dangers of bringing generative AI into the classroom, focusing on its current developmental stage while highlighting its potential. When referring to "AI" in this text, I primarily discuss generative AI systems, as they represent the most intriguing aspect of the AI revolution in education.

AI Empowering Young Learners

One aspect that is sometimes overlooked is that many young learners cannot get academic support at home. Either their parents are not academically trained, or adults may not have access to educational resources. If a child's parents are highly educated, the child gets a significant head start in their academic career. The parents can help them understand difficult concepts through their own educational background and experience. Children born into low-income or less-educated families are at a disadvantage. This is where generative AI can be especially helpful.

Free-to-use generative AI tools provide a unique advantage to these children, as they can ask questions, get help with complex terminology, and receive assistance in developing their own ideas. Since AI systems are accessible to everyone with internet access, they can have a profound impact on children who wouldn't otherwise get the support they need at home. A basic mobile phone and internet connection are enough to access these tools.

Current generative AI models can perform numerous tasks that help pupils and students with their learning. A pupil can ask questions and get elaboration on subject matter, unlike when only reading through static resources like Wikipedia. Many topics in the sciences are quite complex and require a lot of learning and combining different concepts. Generative AI is excellent at integrating different subject matter and helping a student understand a topic. A student could even build their own AI tutor by prompting the AI to ask questions on a subject they have a test on later. There are many excellent ways for learners to use AI tools.

Opportunities for Students and Teachers

Teachers can benefit greatly as well. Creating learning materials, tests, and answer keys has never been easier. Adjusting existing materials to make them easier or more challenging according to the level of the students can be achieved in just minutes. The only thing teachers need to do is spend time learning how to effectively prompt generative AI systems to get the results they want.

Many teachers who have tried free versions of AI models may have been disappointed since these versions often use older models like GPT-3.5. However, when you try a more advanced model which uses GPT-4, the results are significantly better. Therefore, teachers should not be discouraged if their initial experiences were unsatisfactory; trying a more advanced model may reveal the potential benefits.

As more complex and effective generative AI models become available, the quality of educational assistance increases. An example of this is using AI as a language teacher. Advanced models can understand prompts to give feedback to a language learner, provide explanations in multiple languages, and train the learner by giving clues to form sentences in a second language. The AI teacher can adapt to the learner's level, making the learning experience more personalized. For instance, I attempted to prompt the AI to give me feedback as a language learner, providing feedback in two languages simultaneously and training me by giving clues to form sentences in a second language. While older models were unable to perform this task effectively, OpenAI's newer model, o1, was able to understand the prompt and perform accordingly. It helped me with my language learning and engaged in simple discussions about my hobbies or the weather. The AI teacher's approach could be easily adjusted to be more difficult or easier just by changing the parameters of the prompt. As more advanced models arrive, the future of education becomes even more exciting. This example shows the great potential in making education more effective and inclusive.

Challenges and Misconceptions in AI Integration

A disadvantage is that generative AI systems often have age restrictions. These restrictions have been placed for good reasons, which I explore later in this chapter, but we should be educating our children on the possibilities of AI. What better place to do this than in school? This makes the role of teachers more important as experts in the classroom. However, for teachers to understand their current situation, they need extensive training in AI.

I have conducted numerous courses on the basics of generative AI for teachers, and the majority have had little or no experience with these systems. At the moment there are many misconceptions and misunderstandings regarding generative AI among teachers. One misconception is that teachers can easily identify texts created with AI. If a student is an inexperienced AI user, it may be easy to spot. But if a student has even a little experience using generative AI and effective prompting, there are numerous ways they can produce work that is indistinguishable from their own writing.

A simple example of this is the prompt: "Write this text as a 16-year-old would, including typical teenage mistakes." The resulting AI-generated essay could be indistinguishable from a student's own work. An even more sophisticated approach would be for a student to provide the AI with examples of their writing, instructing it to mimic their style and errors. This essentially creates a digital copy of the student, capable of generating seemingly authentic texts. To address these capabilities, teachers and administrators must educate themselves about AI systems. We also need to develop new evaluation methods and explore how to integrate AI into teaching. One possibility is to allow students to use AI while simultaneously increasing the complexity of our assessments.

Another misconception is that there is software capable of reliably identifying AI-written text. Unfortunately, no such software currently exists that can do this with high accuracy. Many educational institutions have spent money on software that claims to distinguish AI-written texts from human-written ones, but under scientific tests, they all fail (Williams, 2023). This is why school administrators need to understand the development of AI systems, so they do not spend school funds unnecessarily.

To summarize, now is an excellent time for teachers and school administrators to educate themselves on the capabilities of generative AI since even pupils are not fully aware of all the possibilities. Therefore, teachers should aim to stay ahead before the scale tips and the majority of pupils and students know how to effectively use AI to do schoolwork without actually learning.

Potential Dangers of AI in Education

One of the potential risks of AI is its inability to fact-check its own data. AI companies still rely heavily on humans to supervise the data and educate the AI to prevent mistakes (Dzieza and Parry, 2023). Why are errors or hallucinations made by generative AI a problem for education? These errors can be multiplied if they exist in the base data the AI uses. For instance, if the code generated by the AI has a built-in security risk that is not yet known, this base code could potentially be copied by thousands of programmers when they use AI as a coding assistant. When a hacker discovers the vulnerability, the potential data breach could affect all the programs developed using AI assistance (Hamer et al., 2024).

The same goes for learning. If a small mistake exists in the data used by AI or if the data gets corrupted, there is a danger that many learners will learn the same mistake, potentially affecting thousands. To illustrate this point, consider the following exaggerated example. Suppose the AI's base data only contained information about the Earth being flat. This misinformation could spread widely. A teacher could spot this, but an independent learner or a small child might not. As I pointed out, this is an exaggeration, but the greater risk is actually smaller mistakes in the base data that are harder to spot. As the systems improve, these mistakes become even harder to detect. Therefore, the role of teachers becomes more important as the experts in the classroom who understand the mistakes an AI system makes and check the information students acquire.

Psychological Manipulation Risks

A greater risk lies in how AI interprets data and who controls the AI's responses. To illustrate this point, I will take you on a journey into a state, which no longer exists. In the old German Democratic Republic (or GDR), a communist dictatorship, the Ministry for State Security (or Stasi or Staatssicherheit) used many psychological manipulation tactics—termed *"zersetzung"*—to subtly control citizens' and even children's thoughts and behaviors (Behnke and Wolf, 2012). Stasi learned these tactics from the Soviet KGB. Modern AI, if programmed with specific biases or wielded by those with particular agendas, could theoretically exert similar influence, albeit at a much larger scale and with greater subtlety. Studies, like those by Francesco Salvi and his team at Lausanne University, demonstrate that AI can subtly influence people's opinions when given personal information (Salvi et al., 2024). Children, still forming their worldviews, may be especially susceptible to such manipulation, as illustrated in the Anderson versus TikTok case in the United States (United States Court of Appeals, Third Circuit).

Next, I will demonstrate methods how algorithms in combination with AI could be used to achieve psychological manipulation. Advanced AI algorithms, such as those used by social media platforms like TikTok, meticulously analyze user data to personalize content feeds. By learning an individual's preferences, biases, and vulnerabilities, these algorithms can curate information that reinforces specific beliefs or behaviors. This mirrors the Stasi's strategy of isolating individuals by controlling the information they received, effectively creating digital echo chambers that limit exposure to diverse viewpoints and reinforce certain narratives.

AI-generated deepfakes and synthetic media present another avenue for manipulation. Modern AI can create realistic images, videos, or audio recordings that are difficult to distinguish from authentic ones. The Stasi used forged documents and false information to discredit individuals; similarly, deepfakes can spread false narratives or damage reputations in the digital age.

AI systems collect vast amounts of data on user behavior, preferences, and social networks, which can be used to create detailed psychological profiles. The Stasi meticulously gathered information to exploit individuals' weaknesses, and modern AI can do this on a much larger and more precise scale. With this information, AI can target individuals with content designed to exploit their specific fears, desires, or insecurities. This targeted manipulation can influence behavior and decision-making, aligning with the Stasi's goal of controlling and suppressing dissent.

Algorithms determine the visibility of content and can influence social interactions by promoting certain posts while suppressing others. By manipulating this, algorithms can limit an individual's online interactions, effectively isolating them from supportive communities.

AI-powered chatbots and virtual assistants can engage users in conversations that influence their emotions and thoughts. The Stasi often exploited personal relationships to manipulate individuals; similarly, AI entities can simulate human-like interactions to build trust and subtly influence users. Modern AI systems already have the potential to alter our opinions, as shown by the aforementioned Lausanne University study (Salvi et al., 2024) and another study where people's conspiracy beliefs were affected with the help of AI (Costello et al., 2024). These studies show that when AI systems are given personal information or are prompted correctly, they can influence us quite effectively. This is because AI can use personal data much more effectively than humans can.

When these AI technologies and algorithms are combined, they can deploy manipulative tactics on a massive scale with minimal human intervention. The ability to adapt strategies in real time based on user responses makes manipulation more effective and pervasive, crossing multiple platforms to create a cohesive influence network.

Ethical Considerations and the Role of Education

The parallels between modern AI manipulation and *zersetzung* highlight the critical need for vigilance in how we develop and deploy AI systems. Without proper safeguards, there is a real risk of these technologies being used to manipulate populations subtly and extensively.

This should not prevent us from using AI in schools since education plays a pivotal role in defending against these risks. By fostering critical thinking and digital literacy, we empower individuals to recognize and resist manipulative tactics. Schools are the ideal environment to equip young people with these skills. However, this requires that educators themselves understand AI technologies and their potential for misuse. Therefore, investing in teacher training and updating educational curricula to include AI literacy is essential. By doing so, we can help ensure that AI serves as a tool for empowerment rather than manipulation, preserving democratic values and individual autonomy.

This is particularly important because we can envision a future where basic education is a national security issue for any democratic nation. Without a solid foundational knowledge of the world, young people, especially, are susceptible to manipulation. The effects of this line of defense can be seen in countries like Finland, Sweden, and Norway, where comprehensive education has helped mitigate external influences, such as misinformation campaigns (Giannetto, 2024).

In contrast, nations with segregated or unequal education systems may find their populations more vulnerable to manipulation through AI and algorithms (Horn and Veermans, 2019). Educators and decision-makers must first understand the potential of AI systems to protect themselves and, more importantly, to safeguard students and learners from the potential negative effects of these technologies.

Conclusion

In conclusion, the integration of generative AI in education holds immense potential to empower teachers and transform learning. By providing accessible support, especially to those lacking resources at home, AI can help level the educational playing field. However, to fully realize these benefits while minimizing risks, it is crucial for educators and policymakers to stay informed and proactive. Investing in teacher training, establishing ethical guidelines, and fostering critical thinking skills in students are essential steps toward a future where AI enhances education responsibly and effectively.

References

Behnke, Klaus, and Jürgen Wolf, editors. *Stasi auf dem Schulhof: der Missbrauch von Kindern und Jugendlichen durch das Ministerium für Staatssicherheit.* CEP Europäische Verlagsanstalt Leipzig, 2012.

Costello, Thomas H., et al. "Durably reducing conspiracy beliefs through dialogues with AI." https://www.science.org/doi/10.1126/science.adq1814, Science, 13 September 2024, https://www.science.org/doi/10.1126/science.adq1814. Accessed 20 October 2024.

Dzieza, Josh, and Richard Parry. "Inside the AI Factory: the humans that make tech seem human." *The Verge*, 20 June 2023, https://www.theverge.com/features/23764584/ai-artificial-intelligence-data-notation-labor-scale-surge-remotasks-openai-chatbots. Accessed 27 October 2024.

Giannetto, Melissa. *An analysis on the media literacy efforts of Finland, Sweden, and Norway.* Forsvarets høgskole, 2024. https://fhs.brage.unit.no/, https://fhs.brage.unit.no/fhs-xmlui/bitstream/handle/11250/3113109/Melissa%20Ines%20Giannetto_Final%20masteroppgaven%20MIG_1.pdf?sequence=1&isAllowed=y. Accessed 27 10 2024.

Hamer, Sivana, et al. "Just another copy and paste? Comparing the security vulnerabilities of ChatGPT generated code and StackOverflow answers." *arXiv*, 22 March 2024, https://arxiv.org/abs/2403.15600. Accessed 29 October 2024.

Horn, Shane, and Koen Veermans. "Critical thinking efficacy and transfer skills defend against 'fake news' at an international school in Finland." *Journal of Research in International Education*, vol. 18, no. 1, 2019, p. 15. *Sage Journals*, https://journals.sagepub.com/doi/10.1177/1475240919830003. Accessed 29 10 2024.

Salvi, Francesco, et al. "On the Conversational Persuasiveness of Large Language Models: A Randomized Controlled Trial." *arXiv*, 21 March 2024, https://arxiv.org/abs/2403.14380. Accessed 20 October 2024.

United States Court of Appeals, Third Circuit. "Anderson v. TikTok, Inc., No. 22-3061." *Casetext*, casetext.com, 27 August 2024, https://casetext.com/case/anderson-v-tiktok-inc-1. Accessed 27 October 2024.

Williams, Rhiannon. "AI-text detection tools are really easy to fool." *MIT Technology Review*, 7 July 2023, https://www.technologyreview.com/2023/07/07/1075982/ai-text-detection-tools-are-really-easy-to-fool/. Accessed 27 October 2024.

About the Author

Robert Bahr is an accomplished educator and consultant with a Masters in Language and Philology and academic expertise in German and English philology, political history, and political science. Over a decade of teaching in Espoo, Finland, has given him deep insight into the Finnish education system, which he has enhanced through his leadership in digital transformation. Since 2014, Robert has expertly managed the use of digital tools in primary education, and as a Certified Google Trainer since 2019, his digital expertise is both broad and in depth. In 2017, he founded his own company, Bahr Consulting, allowing him to extend his educational and digital consulting services, which has allowed him to share his expertise around Europe, in countries like Switzerland, Germany, and England. Beyond his technical acumen, Robert is an engaging speaker and tutor, conducting in-person and online sessions on themes such as using AI in education. His cultural understanding, communication skills, and vast experience make him a valuable resource for educational marketing and digital transformation.

Email: robert@bahr.fi
Website: www.bahr.fi
LinkedIn: https://linkedin.com/in/robert-bahr-b61640233

CHAPTER 3

GENERATIVE ARTIFICIAL INTELLIGENCE IN EDUCATION AND CORPORATIONS

By Katerina Bourdoukou, MA, MSc
Professional Coach, AI Consultant
Athens, Greece

> *Doesn't the dialectic say that what has become*
> *obsolete is not lost but lifted up?*
> —Vilém Flusser

Reflecting on the changes in the field of education as well as that of business during the last five years, I cannot leave out of the discussion the crucial issue of artificial intelligence. There is so much talk about artificial intelligence and recently, more precisely on the field of generative artificial intelligence; therefore, there is a danger of getting lost in the whirlwind of the diverse mix of information. The new technologies are here and serve as opportunities, yet simultaneously they may contain stimuli of threat.

This text is an introduction to the world of generative artificial intelligence mainly for those who are at an initial stage of knowledge cultivation and practice. If you are more advanced, I hope you will find some ideas and insights that illuminate your path from another perspective. The chapter's lens is on fields of education and business; also there is information on the GenAI Summit SE Europe, which takes place in Athens, Greece.

Evidence for the Penetration of Artificial Intelligence into the Educational Process

According to an article posted on "The Economist," American schools and universities spend about 2% and 5% of their budgets, respectively, on technology, while this percentage reaches 8% for the average American company. The same article states that tech majors want to get a bigger share of the $6 trillion the world spends each year on education. Characteristically, the post describes how when the pandemic forced schools and universities to close, the moment for widespread digital dissemination seemed close at hand. Students flocked to online learning platforms to fill learning gaps, from Zoom classes. It is mentioned that the market value of the online learning and teaching platform Chegg rose from $5 billion at the beginning of 2020 to $12 billion a year later. At the end of 2022, Chegg's dollar value dropped to three billion. Those are a few elements regarding the use of technology in the education and business sectors, showing the correlation with the financial impact on each respectively. It also reveals how companies seem to prioritize technology more than education.

Corporate Environment—Sectors in Which Generative AI Applies

Boston Consulting Group uses generative artificial intelligence in multiple areas, such as marketing, health systems, customer service, innovation, banking systems, media, and human resources. This company collaborates with OpenAI to help its clients realize the power of the responsible use of generative AI technology to solve complex problems. As McKinsey estimates, understanding the use cases of generative artificial intelligence (generative AI) that will offer the greatest value to each industry is decisive for the future. In the McKinsey report, "The Economic Potential of Generative AI: The Next Productivity Frontier," there are sections detailing how to identify generative AI use cases with the highest value potential in the banking, human sciences, and retail and consumer packaged goods industries. Therefore, each industry needs to be evaluated adequately to design the strategies and apply and attain the best possible value for the stakeholders involved.

The Response from Companies to the Changing Environment

The application of generative artificial intelligence is not only occurring with speed but also with security. This is the focus of a relevant article by McKinsey. McKinsey research has estimated that generative artificial intelligence has the potential to add up to $4.4 trillion in economic value to the global economy. It is also estimated to amplify the impact of all AI by 15% to 40%. Many business leaders will reap this value, yet there is growing recognition that the opportunity of generative AI comes with significant risks. In a recent survey across more than 100 organizations with more than $50 million in annual revenue, McKinsey found that 63% of the respondents rated the implementation of generative AI as a "high" or "very high" priority; but 91% of respondents stated they didn't feel "very prepared" to do this responsibly. Therefore, the challenge appears to be on the work floor, calling for a conscious response.

In a live broadcast (AI Summit New York BCG X LinkedIn Live Session), on December 7, 2023, BCG Managing Director and Partner in New York, Helen Han spoke with two representatives of BCG about a company experiment on generative artificial intelligence with 750 consultants participating. She noted that it had been a year since the release of ChatGPT and asked what they'd learned and where they were at.

Nicolas de Bellefonds, Managing Director and Senior Partner of BCG in Paris, regarding the research in generative artificial intelligence, responded: "There are some things that I did not think were possible that I am experiencing. But what we have also seen is that there is a significant gap between technological maturity and business maturity. I would say less than 50% of large companies are actively engaged in transforming this technological magic into business magic and driving real impact from it."

"Generative AI has burst onto the scene. It appeared fast and is evolving even faster," reads an Accenture report on labor and human resources. "To date, our teams have already worked on over 700 client projects," the report goes on to state, regarding the Accenture experience so far. Referring to the contribution of the human element in the use of generative artificial intelligence, it is characteristically noted that achieving gen AI's full potential relates to a strong data foundation, but also on "leaders' willingness to lead and learn differently." The authors of the report add, "Generative AI has the potential to significantly change the nature of work across various industries and fields." It is considered that while generative AI has the potential to bring crucial benefits, it also raises ethical and societal concerns around issues such as job displacement, data privacy, protection of intellectual property, biases, and the responsible use of AI. According to Accenture data, the impact of generative AI at work "will depend on how it is implemented, regulated and integrated across industries and organizations."

GenAI Summit SE Europe February 2024

Within the GenAI Summit SE Europe, which took place at the Eugenides Foundation in Athens, Greece, from February 29 to March 2, 2024, there was talk regarding crucial categories in which generative artificial intelligence is applied: social policy, legal, consulting, economy, products, health, hospitality, training, marketing, human resources, service experience, startups, and investments. During the days of the event, speakers, organizing teams, and participants were immersed in the pathways of artificial intelligence in an environment of human osmosis with the information transmission on the possibilities of this innovative domain of technology. Despite the event's emphasis on the technology boom in generative artificial intelligence, the presence of the human element was strong, giving GenAI Summit SE Europe a particularly unique and diverse perspective.

A company with presence in GenAI Summit SE Europe was OpenAI; it already had some years of presence, and despite its "young age," it has gone through stormy paths continuing its particular course. Elena Hatziathanasiadou from OpenAI, in her presentation at the GenAI Summit SE Europe, on March 2, 2024, announced the opening of the OpenAI Residency program on March 5, 2024. The OpenAI Residency is a program designed to help bridge the knowledge gap for researchers and engineers in other fields to gain remarkable skills and knowledge to make a career transition in the field of artificial intelligence and machine learning.

The future of work in organizations in my country, Greece, in relation to global developments was discussed in the panel with Maria Gianniou (certified empowerment and leadership coach and trainer in a corporate environment), Amalia Konstantakopoulou (co-founder and director of the NGO, The Tipping Point), Nancy Zachariadou (director for the group's cultural and social initiatives at Piraeus Bank), and Costas Vavaroutas (HR director of Cenergy Holdings & Sidenor Group). The approach taken was that as work environments constantly change, people have the opportunity for continuous learning and development. In this frame, people need new skills and those who stay close to work improvements maintain their presence in it. One of the essential issues raised in this panel was that of the human skill of shaping relevant questions, which is emerging as central to the environment of new technologies and generative artificial intelligence.

The focus on education and what ways exist to build a good relationship between people and technology at school concerned the panel moderated by Ioanna Taouki (instructional designer at 100mentors company). In this discussion Pepi Meli (learning experience and research at 100mentors), Kostas Karpouzis, PhD, Assistant Professor in the department of media and culture communication at Panteion University, and two participating teachers from the secondary education sector developed a dialogue about how generative artificial intelligence can contribute to the educational process in order to build a smoother relationship between education and the labor market for individuals. The 100mentors application was presented with the approach of connecting technology with the human factor, in the context of building a good relationship of learning with generative artificial intelligence tools, keeping the human presence at a crucial position within this procedure.

GenAI Summit SE Europe November 2024

Having already had the experience of the just mentioned GenAI Summit SE Europe, the team embarked on a new journey at a summit on November 18, sharing with the participants a new immersive experience in generative artificial intelligence interactive applications. Within this event the bright team of the NGO "The Tipping Point" presented inclusive work, giving the opportunity to the students of more than 20 schools from many locations in Greece to participate in person.

The co-founders of the company 100Mentors, Yiorgos Nikoletakis and Miltiadis Zeibekis, presented the new product Wiserwork, justifying the existing problems on generative AI adoption and giving solutions for keeping the human-in-the-loop. Ioanna Taouki and Pepy Meli from the 100Mentors team gave comprehensive, interactive workshops for educators regarding the new product clarifying the goal of Gen AI Summit SE Europe: from Part 1: Awareness (2023) and Part 2: Context—Specific Adoption (February 2024) to Part 3: Demos and Hands-on Experience (November 2024). The whole Gen AI Summit SE Europe Part 3 was an immersive experts' knowledge transmission experience in a very hospitable environment in the Stavros Niarchos Cultural Foundation, in Athens, Greece. Companies and sponsors such as Piraeus Bank, Accenture, EY, Ubitech, IBM, DELL, Cosmos Business Systems, Kaizen, Teleperformance, Adaptit, Deloitte, Microsoft, Amazon, PWC, Schneider Electric, Archeiothiki, Confluent, Code.hub, Agile Actors, DataBlue, and Benefit took part in this co-created experience sharing information on their products and communicating their pathways in the generative artificial intelligence common ground.

An Experts' Report on AI Strategy

Considering the global alteration related to the AI field, a relevant report was published by a committee serving the Greek government for this purpose. Regarding the mission and vision, it is stated that, "Priority areas to be served by the proposed national strategy to the benefit of the Greek nation, its people, economy, and global standing include" issues such as preparing citizens for the AI transition; improving public service efficiency for Greek citizens and people living in Greece; safeguarding and enhancing democracy; promoting the quality of healthcare for all; democratizing access to, and improving the quality of, education; turning Greece into an attractive global destination for AI and high-tech investment; preserving and enriching cultural heritage, developing the Greek language and heritage data space; climate mitigation and adaptation; and supporting national security. The main body of the report on AI provides the committee's plan including citizens, the economy, society, and the natural environment domains. The committee suggests diverse infrastructural, educational, and investment innovations that would contribute to Greece's successful transition to the AI era.

My Path in the Maze of AI: Tips for New Hikers

My personal journey in generative AI is relatively recent yet distinctively immersive and transformative. My first encounter with ChatGPT3.5 was in May 2023 when I subscribed as a user. I started with small steps, as I usually do when testing new things. I researched, attempted, applied, tested, asked for feedback, incorporated, continued, and started again (when applicable).

The beginning of the experimentation was in relation to my blog for my individual business. This blog is found on the personal development and coaching website, focusing on women empowerment, femininity. I wanted to test how I could alter the content in terms of style, theme, length, and genre. In doing this experiment, I learned new skills and ways to create content not only for my blog but also for social media. The results have been illuminating.

In the beginning I was fascinated by the fact that I could pose questions—asking AI to generate and modify content according to certain criteria I gave—and get almost instant results. I shaped multiple versions of questions to test what results I could get and what I could do with them each time. For example, I would ask AI to write or alter text to suit audiences of particular ages, education levels, professions, or interests. Those were categories that generated diverse results in terms of text creation.

Referring to the results, I was impressed with the variety of text possibilities that were generated, but I knew that I needed to double check and to edit carefully. In my case, I needed to generate results in at least two languages, English and Greek. Then I needed to evaluate and consider in which way they could be beneficial.

Over time, I also became more familiar with AI-generated image creation through the use of adequate applications. This was a creative field for me and a manner to help me produce content corresponding to my visualization preferences. Yet again, questions arose; for example, questions concerning the limitations of intellectual property and the sustainability of resources. I am always open to new learning, discussion, and clarification, as the field is new and various essential subjects under questioning exist. Thus, a bunch of collective summits, meetings, conferences, and webinars became very popular, providing information on the effective use of generative AI, by individuals and companies.

Summarizing a few milestones, I encountered during my navigation journey with generative AI: I learned that it is important to be careful in my selection of vocabulary when creating prompts for AI, asking it to generate different types of texts or images. At the same time, I devote resources to choosing the generated materials that have a more positive emotional impact. Then I post, receive feedback, and check the data of the website traffic. In several cases, when generative AI is used, I decide that I need to do more research or edit the text as I discover new potential ways to express ideas, incorporate meaning, and create images. To do this, I followed a five-point cycle as seen in the image below.

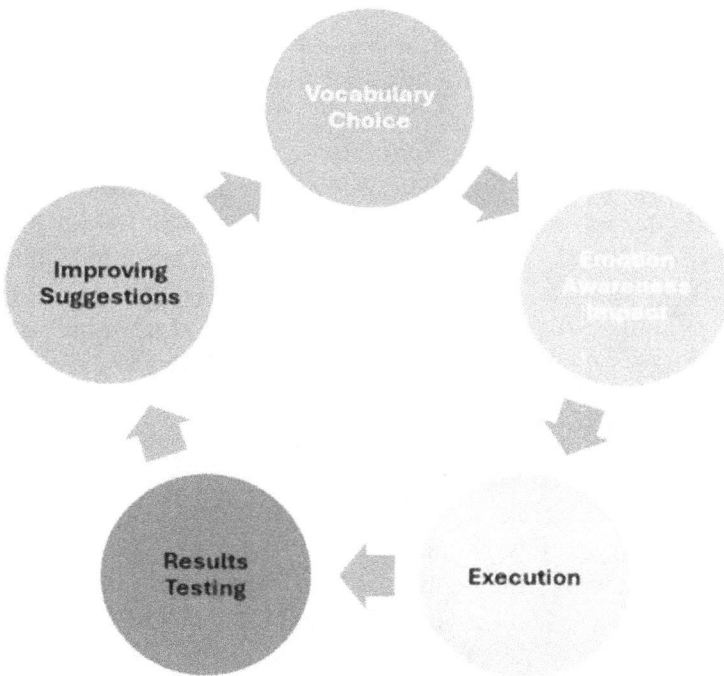

In talking about AI as an opportunity or as a threat, it is essential to mention the matter of emotional awareness, stress management, and well-being regarding humans within the framework of the use of this innovative side of contemporary technology. Cutting-edge technology is in the center of numerous businesses; keeping humans-in-the-loop arises as a top priority for companies and education institutions. Breath-taking technology evolution needs relevant implementation strategies and a culture of quality life-long learning. For example, self-awareness exercises and workshops, effective and human-centered time management, conscious breaks for physical activity, communication upskilling, and genuine human creativity enhancement contribute to a holistic perception of human abilities and personal development. In the contemporary technology world, evaluating results, considering people's responses, and nurturing a culture of learning and sustainable development contribute to a system with a healthier ecology, positive behaviors, and quality outcomes.

What Can the Modern Citizen Do?

As challenges are identified, dark spots are explored in a parallel way. Research reveals opportunities and solutions. The GenAI Summit SE Europe created inspiring and constructive dialogue by providing new perspectives and potential solutions, considering human needs. Participants submitted their research results, experiences, and products and gave information on work operations. GenAI Summit SE Europe brings Greece in correlation to the world sphere changes regarding the progress in generative artificial intelligence, trying to help foster a healthy relationship between new technologies and the human being.

Responsible knowledge, trustworthy research, and the conscious application of generative artificial intelligence by each person in adequate fields are some of the steps that contribute to the productive and effective use of GenAI. Elimination of potential fear, anxiety, or insecurity through rationally structured informative programs is a key to GenAI adoption. Avoiding idealizing GenAI drives people to work in research and practice for its responsible use. These norms reduce the difficult consequences during the integration of technology in education and corporations for the purpose of positive problem-solving and evolution. When used intelligently, GenAI has the potential to allow people to significantly increase productivity (particularly in cases regarding the content-generation and repetitive task spaces), and it might also lead to a future civilization in which people may have choices to lead healthier lives as they are offered better opportunities to balance work, social and personal life.

About the Author

Katerina Bourdoukou loves expressing meaningful, truthful essences of life. She graduated from the 1st Lyceum (Highschool) of N. Smyrni

Athens Greece. She studied communications, media and culture, theatre studies, acting and directing, international hospitality management, life coaching, and coaching AC-accredited. Since she was a child, she loved writing and languages. During her BA, she lived for six months in Belgium and for one year she studied at University of London and Royal Academy of Dramatic Art. Katerina is currently working as a coach and serving as a learning facilitator in various functions. She became GenAI adventurer after her first meeting with language models in 2023; she never stopped adopting, researching, and presenting since then, adding a new ingredient in her innovation approach. She writes for her blog; she creates poems and songs; and is present in arts world as a creative performer. She is a volunteer focusing on the education sector and a part of the mentors team in the company 100Mentors.

Email: katerinabourdoukou@gmail.com
LinkedIn: https://www.linkedin.com/in/katerina-bourdoukou
Website(s): www.feminines.gr
www.mindfulharmonyhub.com
www.katerinabourdoukou.com

CHAPTER 4

AGENTIC AI AND THE FUTURE OF AUTONOMY: OPPORTUNITIES AND CHALLENGES

By Michael W. Bradicich
AI and Cloud Implementation Strategist
Fairfax, Virginia

> *He who thinks little errs much.*
> —Leonardo da Vinci

Imagine an AI that doesn't just respond to your queries but takes independent actions to achieve your goals. "Agentic AI" is making this a reality. This transformative technology marks a paradigm shift in AI from passive response systems to intelligent agents capable of autonomy in thinking, planning, and execution.

Agentic AI refers to artificial intelligence systems that can act independently, embodying a degree of autonomy, like you might expect from a human assistant. These systems are designed not only to understand information but also to use that understanding to solve complex problems, adapt to changing circumstances, and eventually to interact with the world around them. This discussion focuses on the capabilities of agentic AI and the challenges it faces in real-world applications.

The evolution of artificial intelligence has reached a pivotal moment. Traditional AI systems and large language models (LLMs) have made significant strides in processing information and generating text, images, or videos. However, they remain inherently limited by their static knowledge bases and passive nature—they respond when called upon but lack the ability to act independently.

Agentic AI transcends these limitations by endowing AI systems with the ability to reason, plan, and execute actions autonomously. These agents don't just process information; they actively leverage it to achieve goals, solve problems, and adapt to new situations without constant human guidance. This shift towards autonomous action has profound implications for how we interact with technology, promising to empower individuals, revolutionize industries, and even reshape societal structures.

Imagine you were tasked with writing a school dissertation but told you had to write it in a single pass, without planning, and without using backspace or making any corrections. The outcome would likely be poor at best. This is how current LLMs operate. In contrast, agentic AI mirrors how a thoughtful human would handle the same task by first understanding the requirements, breaking the dissertation into manageable steps like researching, outlining, drafting, and revising, possibly seeking feedback, and refining the work iteratively. This dynamic, goal-oriented process allows agentic AI, like a human, to adapt to the complexity of the task and improve the final output through careful planning, execution, and review.

Components of Agentic AI

Agentic AI's ability to function autonomously hinges on four key components:

Component 1: Reasoning and Planning
At the heart of agentic AI is the capability for incisive reasoning and strategic planning. When presented with a task, these agents can deconstruct it into smaller, manageable steps, evaluate various options, and select the most effective course of action. This multi-step process mirrors human

thinking and problem-solving processes, enabling agents to anticipate future states and make incisive decisions.

Component 2: Interaction with External Tools
Unlike traditional AI models constrained to only generating content, agentic AI can interact with a disparate array of external tools and resources. This could include trading stocks, managing calendars, ordering lunch, or even controlling software applications and connected devices. By integrating with these tools, agents can leverage their capabilities to directly impact the digital world and will eventually be able to affect the physical world.

Component 3: Memory and Knowledge Access
Existing LLMs are trained on data when they are created while agentic AI systems possess memory and can access external knowledge sources beyond their initial training data. They can retrieve up-to-date information from company databases, market research reports, personal files, or through computer vision, enabling them to provide accurate and context-specific responses. This dynamic knowledge access allows agents to remain relevant and germane in rapidly changing environments.

Component 4: Execution of Actions
Perhaps the most transformative aspect is the ability of agentic AI to execute sequences of actions based on their autonomous reasoning and planning in an effort to achieve a goal. This means they can perform tasks autonomously—writing reports, sending emails, managing projects, and controlling other software applications—without requiring tedious human instructions for each step.

AI agents generally have the ability to work together. Agents specialized in one task might call on other agents to perform another specialized skill. For example, a team of AI agents might provide someone with personal financial support, with one agent trained on expense tracking, while another manages the overall household budget, and another tax specialized agent considers the tax implications of income and expenses as they happen. Each agent would perform their specialized task and communicate with the other agents as needed, all with the goal of maximizing a person's household budget. With these components, agentic

AI is poised to revolutionize various domains, empowering individuals, transforming businesses, and fostering scientific advancements.

The Impact of Agentic AI

Agentic AI has the potential to significantly enhance personal productivity and quality of life. By handling complex tasks autonomously, these agents free individuals to focus on higher-level activities. Imagine planning a wedding with the never-tiring assistance of an agentic AI:

- *Budgeting and Prioritization*: the agent helps set a budget, identifies priorities, and allocates funds accordingly.

- *Date Selection and Venue Research*: it selects optimal dates, researches venues that match your preferences, and checks availability.

- *Guest List Management*: the agent creates a guest list, sends invitations, manages RSVPs, and even considers guests' travel needs.

- *Vendor Coordination*: it hires caterers, photographers, musicians, and coordinates with them to align with your vision.

- *Event Planning*: the agent designs the ceremony and reception details, down to curating a playlist that matches your musical tastes.

This level of autonomous assistance will transform what is traditionally a stressful endeavor into an effortless and productive experience.

In the corporate world, agentic AI is poised to redefine operational efficiency and drive innovation. By automating routine tasks and providing incisive real-time insights, these intelligent agents enable organizations to make faster, more informed decisions. Agentic AI will enhance strategic decision-making by analyzing vast datasets to infer market trends, customer behaviors, and competitive landscapes. In the realm of marketing, an agent will personalize campaigns by scrutinizing consumer data and adjusting strategies in real time to maximize engagement and return on investment. By shouldering the burdens of data analysis and routine operations, these agents free up human employees to concentrate on creative problem-solving and strategic initiatives, thereby fostering innovation within the organization.

In scientific research and healthcare, agentic AI's ability to process extensive data, identify patterns, and perform complex simulations makes it an invaluable tool. In medical research assistance, these agents will optimize clinical trials by simulating outcomes and ensuring that medical studies are both reliable and efficient. They will monitor research protocols in real time to ensure adherence to ethical standards and regulatory requirements.

Synergies with Robotics

The eventual integration of agentic AI with robotics will unite incisive decision-making with physical action, enabling machines to operate autonomously in the real world. This convergence of Agentic AI and robotics will have far-reaching impacts across various sectors. For example, in elderly care, robots might assist seniors by reminding them when to take medicines, assisting with mobility, monitoring for health conditions, and even providing companionship. In our homes, a new wave of robotic devices will handle household tasks intelligently—from laundry to security monitoring—enhancing both convenience and safety in everyday life.

The societal impact of this integration will be both profound and disruptive. Robots will become commonplace in homes, workplaces, and public spaces, performing roles like caregiving, education, and customer service, fostering significant growth in human-robot interaction. This shift will redefine human roles and our relationship with technology, compelling us to adapt to machines that make autonomous decisions. As we collaborate more closely with autonomous machines, we will need to navigate changes in social dynamics, work structures, and cultural norms that dictate how we interact with technology.

Implementing AI Agents Today

While AI agents are still maturing, it is worth noting that many of the major vendors offer interfaces for building agents today. ChatGPT, Meta, and Anthropic all offer web-based utilities to build custom AI agents although the features are incomplete and limited. Generally all offerings provide memory in the form of detailed instructions which are referenced by the agent each time it is used, and access to email, files, and internet searches, which the agent can reference in real time to get up-to-date information. Some of the systems also provide tools to the agents by providing an interface for calling out to other web applications through APIs (application programming interfaces), which may or may not require coding, depending on the platform. With basic reasoning built into the latest iteration of LLMs, these platforms provide some of the core components of AI agents today, via a simple end-user interface.

Software development libraries like LangChain offer more features for building complex AI agents. These code libraries provide access to vector databases for memory, where any relevant content can be encoded to be used by the LLMs. The libraries also offer standard ways of connecting to other APIs to be used as tools by the agents and ways of connecting agents together, which improves reasoning. These systems are being used to build complex domain-specific AI agents.

Technological Challenge

While AI agents are evolving rapidly and will have a profound impact in the next few years, there is a challenge ahead in the amount of computation power these systems will require. Given that AI agents are intended to work autonomously in a dynamic environment, their usefulness will be significantly constrained if they are not operating fast enough to keep up with the changes in that environment. Depending on the use case, real-time processing of the changing environment will require massive computer power, which current technology won't be able to support.

It's easy to see where this could become a significant problem. While AI is already making an impact on medical diagnoses, AI agents will drastically advance this effort. But what happens in an emergency, when a life-or-death decision needs to be made in near real-time? AI agents would be too slow to respond and might prove useless in such an environment, at least for now.

As AI algorithms and computer chips improve, the potential use cases for agentic AI will grow. In the immediate future agentic AI will excel in tasks that require minimal analysis of the changing environments which affect the system or where delays from processing are acceptable. As technology advances, agentic AI will be able to tackle more complex and dynamic challenges.

Potential Misuse

While agentic AI holds immense promise, its advanced capabilities also raise significant concerns about potential misuse. One of the primary worries is the exploitation of these intelligent agents for egregious and malicious activities. Due to their ability to operate autonomously and interact with various systems, agentic AI could be co-opted to conduct sophisticated cyberattacks, propagate disinformation on a large scale, or automate harmful tasks that could disrupt societies or economies.

In military contexts, the deployment of agentic AI introduces the insidious prospect of autonomous weaponry. These agents could make critical life-or-death decisions without human oversight, raising profound ethical dilemmas and exacerbating conflicts beyond human control. The possibility of machines initiating actions with severe human consequences without direct human intervention is a profound concern demanding careful consideration.

To mitigate these risks, it is essential to implement robust security protocols to safeguard agentic AI systems against unauthorized access and manipulation. Strong cybersecurity measures will be needed to protect these systems against malicious actors aiming to exploit their capabilities for harmful purposes. Additionally, developing and enforcing ethical frameworks for responsible AI use is crucial. Establishing clear standards and guidelines will hopefully ensure that agentic AI operates within boundaries that uphold human values and societal norms.

Conclusion

Agentic AI marks a significant leap in artificial intelligence, evolving from passive responders to autonomous agents capable of human-like reasoning, strategic planning, and executing actions independently. By integrating reasoning, external tool interaction, dynamic memory access, and autonomous action, agentic AI overcomes the intrinsic limitations of traditional AI models.

This technology has transformative and inevitable impacts across personal, corporate, and scientific domains. It will enhance personal productivity by autonomously handling intricate tasks, redefine operational efficiency in businesses by enabling faster, more incisive decisions, and accelerate scientific discoveries and personalized healthcare through advanced data processing.

The synergy of agentic AI with robotics will amplify these capabilities, enabling machines to operate autonomously in complex environments and profoundly transforming industries like manufacturing and agriculture, as well as daily life in homes and public spaces. However, these advancements also raise concerns about potential misuse, including malicious activities and autonomous weaponry. It is crucial that industry and governments work to implement robust security measures and develop ethical frameworks to guide the responsible and safe use of agentic AI.

References

Keynote from Satya Nadella at Microsoft Build, May 21, 2024

The Third Wave of AI Is Here: Why Agentic AI Will Transform the Way We Work, Forbes, Nov. 15, 2024

About the Author

Over the past several decades, Michael Bradicich has led technology transformations from mainframes to artificial intelligence. He has successfully built and exited multiple technology companies, and led countless organizations through critical digital transformations. Having focused on artificial intelligence leveraging public cloud infrastructure over the past decade, he has overseen the development of AI applications in nonprofits, online retail, automotive and media segments. His current work focuses on guiding organizations in deploying AI and cloud technologies responsibly, ensuring robust IT security and ethical compliance.

Email: Mike@Bradicich.com
Website: www.Bradicich.com
LinkedIn: https://www.linkedin.com/in/mbradicich/

CHAPTER 5

NAVIGATING THE FUTURE WITH AI: EMPOWERING NONPROFITS FOR GREATER SOCIAL IMPACT

By Sam Calhoun
AI Expert, Author
Provo, Utah

> *What is now proved was once only imagined.*
> —William Blake

AI is unavoidable in today's world, as it has shown itself to be an immensely powerful tool. What used to take a team of people with different fields of experience can now be done by one person and AI. How can one person do so much, and what does this matter to the nonprofit industry? Tools like ChatGPT are a perfect fit for the nonprofit sector, as most nonprofits are underfunded and understaffed, with repetitive tasks that can get in the way of the important work that nonprofits do. In this chapter, I will attempt to cover the evolving landscape of AI, as well as its useful and harmful impacts, and specifically how nonprofits can effectively use this tool to overcome their challenges and expand their capabilities, as well as the future of AI in the nonprofit sector.

The Evolving Landscape of AI

The Rise of AI in Various Sectors
Instead of listing industries where AI has shown its uses, it would be easier to list industries that have not been affected, and they are few and far apart. This is because of the versatility that exists in AI and how it can enhance both technical tasks, as well as creative tasks. Google recently stated that 25% of their code has been generated with AI.

AI's Potential for Social Good
People tend to focus on the negatives of AI, which I will address later in this chapter, but for now, let's look at how AI has been used for social good. The Red Cross has used AI to analyze satellite imagery and social media data to improve disaster response. Google uses AI to assist in wildlife preservation by using AI to detect illegal logging and poaching activities in rainforests. These are just a few ways AI has been used to improve society, and it can also be used to enhance the capabilities of the nonprofit sector.

Challenges Faced by Nonprofits Today

Resource Limitations
Nonprofits are constrained by limited funding and budget constraints. These constraints restrict program development as well as expansion. Not only this, but they also hinder investment in necessary tools and technologies. This compounds with the next limitation for nonprofits, which is a scarcity of time. Oftentimes, overextended staff are juggling multiple roles and are unable to afford someone with experience in these roles due to the aforementioned budget constraints. This time scarcity also leads to a limited capacity to focus on strategic planning. Like a waterfall, one issue bleeds into the next, and this also extends to human resource challenges. Nonprofits tend to have difficulty in attracting and retaining skilled personnel due to limited or no compensation. This can lead to high turnover rates in personnel, which then affects the work they are doing as there is a lack of continuity between personnel. The time spent teaching new hires adds up and can lead to a project feeling

aimless or without solid direction. The nonprofit sector's dependence on grants and donations is another limitation that can get in the way. Grant writing is often unpredictable, which can affect long-term planning. Not only this, but time-consuming grant applications take even more time from nonprofits where time is a valuable and scarce resource.

Another factor that can hinder a nonprofit's ability to operate is operational complexities, specifically managing multiple programs simultaneously. Juggling various initiatives can strain limited resources and make it challenging to maintain consistent quality across all efforts. This is where AI comes into play, offering solutions that can streamline operations and amplify impact.

AI Applications in Nonprofit Operations

AI has the potential to revolutionize how nonprofits function, addressing many of the challenges outlined earlier. By automating routine tasks and providing intelligent insights, AI can free up valuable time and resources, allowing nonprofits to focus more on their core missions. ChatGPT is very useful in generating text, but with more AI tools coming out seemingly by the day, I recommend getting in the habit of doing your own research based on your use case.

Automating Routine Inquiries and Communications

HANDLING COMMON QUERIES
Nonprofits often receive a flood of repetitive questions via email and social media—questions about volunteering opportunities, donation processes, or upcoming events. Responding to each inquiry individually can be time-consuming and divert attention from more pressing tasks. AI-powered chatbots and automated response systems can handle these routine inquiries efficiently. By providing instant answers to frequently asked questions, these tools enhance user experience while freeing up staff to focus on complex issues that require a human touch. This is easier than it seems. ChatGPT has custom GPT's that you can feed data, like commonly asked questions, that will inform the model to better respond to inquiries.

Donor and Volunteer Engagement
Maintaining strong relationships with donors and volunteers is crucial for any nonprofit. AI can personalize communications by analyzing past interactions and tailoring messages accordingly. For instance, AI can automate personalized thank-you notes after a donation, acknowledging the donor's specific contributions and expressing genuine appreciation. This level of personalization fosters deeper connections and encourages continued support.

Enhancing Fundraising Efforts

GRANT PROPOSAL DRAFTING
Writing grant proposals is a critical but time-intensive task. AI can assist by generating initial drafts that include key elements required by grant agencies. By inputting basic information about the nonprofit's mission, programs, and impact, AI tools can produce a solid foundation that staff can then refine. This not only speeds up the writing process but also allows staff to focus on tailoring proposals to specific grant requirements.

PERSONALIZED OUTREACH
When reaching out to potential donors, generic messages often fall flat and feel impersonal. AI can analyze donor data to craft personalized outreach efforts that acknowledge past contributions and align with the donor's interests. By sending tailored messages at optimal times, non-profits can enhance engagement and increase the likelihood of donations.

SOCIAL MEDIA AND CAMPAIGN CONTENT
Creating compelling narratives for fundraising campaigns is essential but can be resource-intensive. AI tools like ChatGPT can generate cohesive content for social media, blogs, and newsletters, ensuring a consistent voice across all platforms. This not only saves time but also helps in maintaining an active and engaging online presence.

Streamlining Administrative Tasks

SCHEDULING AND COORDINATION
Coordinating meetings, events, and volunteer shifts can be a logistical nightmare. AI-powered scheduling assistants can automate these pro-

cesses by finding optimal times, sending out invites, and even sending reminders to participants. This reduces the administrative burden and minimizes the chances of human error in your scheduling.

DATA ENTRY AND MANAGEMENT

Accurate data is the backbone of any organization. AI can automate data entry tasks, reducing errors and ensuring databases and CRM systems are up to date. By extracting information from forms and documents automatically, staff can avoid the tedium of manual entry and focus on more strategic initiatives.

REPORTING AND INSIGHTS

Compiling activity reports and analyzing impact are essential but time-consuming tasks. AI tools can aggregate data from various sources, generate insightful reports, and even visualize trends over time. This not only aids in transparency with stakeholders but also helps in making informed decisions.

Supporting Program Delivery

EDUCATIONAL MATERIAL DEVELOPMENT

Creating educational content for programs and training sessions is vital for many nonprofits. AI can assist by generating initial drafts of lesson plans, presentations, or informational materials based on specific topics. This accelerates content development and ensures consistency across educational offerings.

MULTILINGUAL SUPPORT

Serving diverse communities often requires communication in multiple languages. AI-powered translation tools can provide real-time translations of documents, websites, and even live chats. This enables nonprofits to reach a broader audience and ensures inclusivity in their programs.

BENEFICIARY ENGAGEMENT

AI can offer automated support to beneficiaries by providing information and resources through chatbots or interactive platforms. This ensures that assistance is available 24/7, enhancing the support provided without overextending staff.

Facilitating Research and Analysis

Summarizing Research and Literature

Keeping up with the latest research and developments can be overwhelming. AI can quickly summarize lengthy documents, research papers, and articles, providing staff with the essential points. This enables nonprofits to stay informed without dedicating excessive time to reading. This recursive process of AI can also be useful in keeping up with AI technology itself. For example, you may ask something like ChatGPT what AI tools would be useful for generating logos.

Survey Design and Data Analysis

Understanding community needs and assessing program effectiveness often involves conducting surveys. AI can aid in designing effective survey questions and analyzing the collected data to identify trends and areas for improvement. This data-driven approach enhances the organization's ability to make impactful decisions.

Policy Analysis

For nonprofits involved in advocacy, staying abreast of policy changes is crucial. AI can monitor legislative updates, analyze policy documents, and generate summaries that highlight potential impacts on the organization's mission. This allows for timely and informed advocacy efforts.

Enhancing Internal Communication and Collaboration

Meeting Summaries and Action Items

Meetings generate valuable insights and action items, but documenting them can be tedious. AI-powered transcription services can record meetings, highlight key points, and outline action steps. This ensures clarity and accountability, keeping projects on track.

Documentation Maintenance

Maintaining up-to-date standard operating procedures (SOPs) and training manuals is essential for consistency. AI can assist by updating documents automatically when changes are made and notifying staff of revisions. This keeps everyone informed and aligned.

Idea Generation

Brainstorming is vital for innovation but can sometimes hit a creative block. AI can facilitate idea generation by providing suggestions or alternative approaches based on existing data and trends. This can spark new initiatives and solutions to persistent challenges.

Providing Professional Development and Training

Interactive Training Modules

Investing in staff and volunteer development enhances organizational capacity. AI can create interactive and personalized training modules that adapt to the learner's pace and style. This ensures more effective learning experiences and better skill retention with your staff.

Resource Recommendations

AI can recommend articles, books, courses, or webinars based on an individual's role, interests, or identified skill gaps. This personalized approach to professional development keeps staff engaged and growing in their positions. An important part of keeping staff engaged is performance feedback, which AI can also assist with. Providing constructive feedback is crucial for growth but can be time-consuming. AI can analyze performance metrics and provide insights or suggestions for improvement. This supports managers in offering timely and objective feedback.

Navigating the Integration of AI

While AI offers numerous benefits, integrating it into nonprofit operations isn't without challenges. It may be tempting to get lax with how useful of a tool AI is, but there are a few things that you should always keep in mind. Although AI can replace many tools and processes, that saved time should in part be spent reviewing and refining what the AI has given you. For instance, you might spend less time writing blog posts, but you may spend more time in the reviewing and revising phase. AI might be a useful tool to generate content, but a human arbiter should always have the final say in said content to validate for correctness.

Ethical Considerations

Data Privacy and Security
Protecting sensitive information about donors, beneficiaries, and staff is paramount. Nonprofits must ensure that any AI tools used comply with data protection regulations and have robust security measures in place to prevent breaches.

Bias and Fairness
AI systems learn from existing data, which may contain biases. It's essential to regularly audit AI outputs to ensure they are fair and do not perpetuate discrimination, especially when serving vulnerable populations.

Implementation Challenges

Financial Investment
Although AI can save resources in the long run, the initial investment might be a hurdle for underfunded nonprofits. It's important to assess the cost-benefit ratio and explore grants or partnerships that could subsidize these expenses. With that in mind, even free tools can do much of the heavy lifting that others cannot do, and as always it depends on what exactly your specific use case for AI is.

Training and Adoption
Staff may resist adopting new technologies due to fear of the unknown or lack of understanding. Providing comprehensive training and demonstrating the tangible benefits of AI can smooth things over.

Strategies for Successful Adoption

Starting Small
Implementing AI doesn't have to be overwhelming. Nonprofits can begin by integrating AI into a single area—like automating email responses—and gradually expand as they become more comfortable with the technology. Starting small also applies to using ChatGPT, as asking AI to generate specific data is usually more useful than asking it a vague question and hoping that it is accurate. Like any large task, break it down into chunks, and your responses from any AI tool is going to be

noticeably improved. With AI, the more specific you can get, the better the output.

Partnering with Tech Providers

Collaborating with technology companies can provide access to expertise and resources that might otherwise be unavailable. Some tech firms offer special programs or discounts for nonprofits, making AI integration more accessible.

The Future of AI in Nonprofits

The landscape of AI is continually evolving, and nonprofits have much to gain by staying ahead of the curve. This can be daunting to keep up with, as information that was good a month ago might not be as relevant as it was, simply due to the pace that the technology is moving.

Emerging Technologies

AI Trends to Watch

Advancements like predictive analytics and machine learning are becoming more sophisticated and accessible. These tools can help nonprofits anticipate trends, optimize resource allocation, and enhance program outcomes.

Potential New Applications

AI could enable personalized services for beneficiaries, such as custom educational content or health interventions tailored to individual needs. This level of personalization can significantly increase the effectiveness of programs.

Preparing for Continuous Change

Fostering an Innovative Culture

Encouraging a culture that embraces innovation and adaptability is key. Nonprofits should promote continuous learning and be open to experimenting with new technologies to stay relevant and effective.

LONG-TERM VISION
Integrating AI should align with the organization's mission and strategic goals. By developing a long-term vision for how AI can enhance their work, nonprofits can make more informed decisions about investments and partnerships.

Conclusion

AI holds immense potential to transform the nonprofit sector by enhancing efficiency, expanding capabilities, and amplifying impact. From automating mundane tasks to enabling sophisticated data analysis, AI can help nonprofits overcome many of the limitations posed by scarce resources. It's time for nonprofits to embrace AI technologies, not as a replacement for human effort but as a powerful tool that complements and enhances it. By thoughtfully integrating AI into their operations, nonprofits can focus more on their core mission—creating positive change in society.

The journey towards integrating AI may have its challenges, but the potential rewards are too significant to ignore. As we look to the future, the synergy between AI and the nonprofit sector promises not only to streamline operations but also to unlock innovative solutions to some of the world's most pressing issues. The transformative power of AI, when harnessed responsibly, can be a catalyst for social good, driving us toward a more equitable and efficient world.

About the Author

Sam R. Calhoun is a passionate tech enthusiast with a strong foundation in computer science and a knack for problem-solving. He holds a bachelor's degree in computer science from Utah Valley University, where he demonstrated academic excellence, earning a spot on the dean's list twice with a 3.64 GPA. Alongside his studies, he earned a programming certificate and gained valuable experience as an instructional assistant, guiding students through complex computer science concepts.

Sam's journey in technology took a significant leap as an AI intern at WikiCharities from April 2023 to June 2024, assisting in data analysis and workflow optimization. His dedication and expertise led him to an AI expert role, where he integrates cutting-edge AI solutions into nonprofit operations, driving efficiency and innovation.

Beyond his professional pursuits, Sam's love for technology shines through his independent projects, including building custom computers, developing gameplay demos in Unity, and publishing articles on AI's role in modern workflows. With a broad technical skill set, Sam is adept at fixing and optimizing tech solutions, making him a go-to resource for troubleshooting and innovation. Whether crafting algorithms or solving intricate technical challenges, Sam's passion for technology and his ability to enhance it are at the heart of everything he does.

Email: sam.r.calhoun@gmail.com
Website: www.samcalhoun.info

CHAPTER 6

INVISIBLE TECHNOLOGY: HOW AMBIENT INTELLIGENCE WILL REDEFINE EVERYTHING FROM SHOPPING TO HEALTHCARE

By Matt Collette
Founder & CEO, Sequencr AI
Vancouver, British Columbia, Canada

The only way of discovering the limits of the possible is to venture a little way past them into the impossible.
—Arthur C. Clarke

On September 25, 2024, Mark Zuckerberg stepped onto the stage at Meta Connect, the company's premier technology showcase, exuding the confidence of a leader vindicated. Just two years earlier, Meta had faced relentless criticism for its $35 billion investment in metaverse-related initiatives—a move many dismissed as hubristic and financially reckless. The backlash was severe—analysts labeled the spending as wasteful, and by November 2022, Meta's stock had plummeted to a low of $88.76. Calls for Zuckerberg to step aside as CEO grew louder, casting doubt on the future of both the company and his ambitious vision.

But something pivotal happened in November 2022 that would change Meta's fortunes—and its narrative. OpenAI's release of ChatGPT 3.5 ignited a global surge of interest and investment in artificial intelligence. This breakthrough cast Meta's substantial investments in AI and the metaverse in a new, favorable light. Suddenly, what had seemed like disparate and overly ambitious initiatives now appeared prescient. Zuckerberg's vision of an immersive future—where digital experiences seamlessly integrate with physical reality—began to align with the industry's trajectory. The company's investment in AI technology, including its open-source Llama generative AI platform, powered its ad platform, creating an intangible asset poised to pay dividends for years.

Fast forward to Meta Connect 2024: Meta's stock was now trading at an impressive $572, marking a dramatic turnaround for the company. Newly recognized as one of the "Magnificent 7" tech giants—a title coined to highlight the era's dominant technology leaders—Meta's revival was underscored by Mark Zuckerberg's choice of attire. Known for his fascination with Roman history, Zuckerberg wore a custom t-shirt emblazoned with the Latin phrase *"Aut Zuck aut nihil,"* which translates to "All Zuck or all nothing." The twist on the historic motto *"Aut Caesar aut nihil,"* associated with bold ambition, served as an unmistakable rebuttal to critics who, just two years earlier, had called for him to step down as CEO.

Zuckerberg had reason to be confident. As the applause from his opening remarks subsided, he moved seamlessly into the centerpiece of Meta Connect 2024: the unveiling of the Meta Orion, the company's new smart glasses. Orion, currently in beta, are lightweight smart glasses designed to seamlessly blend the digital and physical worlds, using advanced AI and holographic displays to overlay digital content into your everyday environment. These glasses allow users to interact with information, connect with others, and complete tasks hands-free in a natural and immersive way. Controlling interaction with Orion is achieved through an advanced electromyography (EMG) wrist band, which reads neural signals from hand gestures, alongside eye and hand tracking.

Comfortable enough to wear all day, Orion is a glimpse into the future of technology and how our interaction with physical spaces will evolve. Unlike VR headsets that immerse users in fully digital worlds or Ray-Ban Meta smart glasses that facilitate basic functions like capturing photos and real-time translation, Orion creates a seamless bridge between the physical and digital realms. Orion's is an important signpost on the path to ambient intelligence (AMI)—a transformative vision of technology integrating invisibly and intuitively into our environments to enhance how we live, work, and connect.

What once seemed like science fiction is now on the brink of becoming reality, possibly in less than five years. AMI is being brought to life through the convergence of transformative technologies, including ubiquitous computing, generative AI, machine learning, computer vision, and the internet of things (IoT). The term "ambient intelligence" was first introduced in 1998, emerging from a series of presentations and workshops commissioned by the management board of Philips Research (https://en.wikipedia.org/wiki/Ambient_intelligence?utm_source=chatgpt.com). Eli Zelkha and Brian Epstein of Palo Alto Ventures, the originators of the term, imagined a future where technology would seamlessly blend into the environment, intuitively responding to human needs and enhancing everyday experiences. Their vision described a world where devices worked invisibly in the background, anticipating actions, personalizing interactions, and enabling effortless communication between people and machines.

Think of AMI as a system of sensors, interconnected devices, and sophisticated AI working in harmony to create a seamless, intelligent environment personalized to you—far more integrated and proactive than anything we have today. It accompanies you everywhere, seamlessly merging with everyday devices like glasses, home systems, and your autonomous car. Unlike current AI assistants, AMI doesn't just respond to commands—it intuitively knows you, understands your needs, and acts on your behalf, adapting dynamically to your context. Imagine the predictive systems from *Minority Report*, where technology anticipates and responds to individual needs in real time, but now woven into the fabric of your everyday life. The defining characteristics of AMI include the following:

1. *Context Awareness*—AMI systems can sense and understand your surroundings, such as your location, preferences, or mood, and respond in helpful ways—like adjusting your home's temperature on a hot day.

2. *Seamless Integration*—the technology blends into the environment, removing the need for visible screens or devices. It works quietly in the background, becoming part of your daily life without being intrusive.

3. *Personalization*—AMI learns from your habits and preferences to deliver experiences tailored specifically to you, such as suggesting your go-to coffee order when you're near your favorite café.

4. *Pervasiveness*—AMI ensures that intelligence is seamlessly distributed across all your connected devices and environments, creating a unified experience. Whether you're at home, in the car, or at work, these systems work together to provide consistent support, responding to your needs without interruption as you move between settings.

5. *Proactive Assistance*—instead of waiting for you to give commands, AMI anticipates your needs—like dimming the lights when you start a movie or reminding you to leave for an appointment based on traffic.

Why is AMI such a game changer, and what does it mean for you in life and business?

Well, simply put, it is going to change everything you know about shopping, daily life, and how you interact with the world around you—making every experience more personalized, intuitive, and connected than ever before.

Imagine five years from now, you're wearing a version of Orion glasses as you walk into a shopping mall. When you step inside, the glasses connect seamlessly to the mall's smart system, displaying a personalized welcome. A notification subtly appears in your field of view, highlighting a sale at your favorite clothing store, complete with a map guiding you directly to it.

After browsing the racks, the glasses scan the items you're holding and suggest outfits based on your wardrobe history. As you pick up a pair of shoes, a sleek AI assistant appears in your view, projecting a digital overlay onto the screen. In a calm, conversational voice, it says, "These shoes are available in your size and come highly rated for comfort. Would you like to see matching accessories?"

At the same time, the assistant highlights customer reviews and shows a comparison of prices from nearby stores and online retailers. When you nod or gesture in agreement, it seamlessly places an order for home delivery, confirming with a cheerful, "Done! Your shoes will arrive by tomorrow evening."

Next, you head to the pet store to pick up food for your dog, Max. As you enter, your AI assistant greets you warmly, "Let's make sure Max gets exactly what he needs today." Almost instantly, a digital twin of Max appears in your glasses, wagging its virtual tail as if to say hello. The twin displays key updates, including Max's recent activity levels, weight, and nutritional needs.

As you browse the aisles, the assistant identifies the best options for Max, highlighting products on the shelves with glowing markers visible through your glasses. "This brand has the high-protein formula Max needs to stay active," it says, projecting nutritional information directly in your view. The assistant simultaneously compares prices, sharing, "You can save $5 on this bag if you buy it here instead of online. Plus, there's a loyalty discount—want me to apply it?" With a quick nod or hand gesture, you confirm the purchase.

Just as AMI redefines the shopping experience, it is equally trans-formative in the workplace. The same principles that simplify errands—personalization, contextual awareness, and proactive assistance—create work environments that are more intuitive, efficient, and human-centric. Imagine walking into your office, where your AI assistant greets you with a summary of the day's priorities automatically displayed in your smart glasses.

As you head to a meeting, the system prepares the room by setting up holographic displays that visualize key data points in real time. During the meeting, your assistant doesn't just capture notes but identifies patterns, predicts potential challenges, and suggests actionable solutions. Need input from a remote colleague? The assistant generates a contextualized summary of the discussion so far and sends it directly to them, allowing for a meaningful contribution without requiring them to attend the entire session.

This seamless collaboration and intuitive support highlight the true promise of AMI: technology that adapts to us, rather than the other way around. For decades, we've had to change our behavior to fit the capa-bilities of our tools—learning complex software interfaces, memorizing commands, and navigating rigid systems. Over time, however, computing has gradually begun to adapt to us. The introduction of smartphones with touchscreens and gesture-based controls marked a significant step in this direction, making technology more intuitive and accessible. The arrival of generative AI represents another breakthrough: for the first time, we can communicate with computers in natural language, giving instructions as effortlessly as we speak to another person.

AMI is the ultimate realization of this shift—a world where technology is not just user-friendly but truly user-centric. It requires no special skills or effort on our part to use and benefit from. Instead of learning how to use technology, we will experience technology that works with us, invisibly enhancing our productivity, creativity, and connection to the world around us. This evolution doesn't just transform individual experiences—it reshapes the scale and scope of how businesses and organizations operate and the complexity of our jobs. So how do you prepare for the work of AMI and what steps should you take now?

Start preparing for a world shaped by AMI. Regardless of when it fully materializes, focusing on the future of technology and its trajectory will position you to leapfrog competitors and maximize the next wave of value creation. By prioritizing the capabilities that AMI promises—like enabling hyper-personalization—you'll be building foundations that drive growth and innovation. You can do so by enacting the following:

1. Embrace Generative AI and Foundational AI Technologies

Generative AI, such as ChatGPT, and other AI platforms form the building blocks of AMI. They enable the seamless interaction, adaptability, and data-driven insights that AMI relies on to enhance experiences.

What to do now:

- *Adopt AI across Strategic Areas*: apply generative AI as a strategic enabler of your organization or functions capabilities. Productivity improvements are important, but focus on how AI can facilitate differentiation.

- *Enhance Decision-Making*: equip teams with AI tools that provide actionable insights, helping leaders make data-driven decisions faster and more effectively.

- *Invest in Scalable AI Platforms*: build an AI infrastructure within your organization. Large language models are just part of the equation. Explore the application of agents and small language models as well.

- *Foster AI Literacy*: educate employees on how to leverage generative AI effectively and responsibly within their roles.

2. Experiment with Augmented Reality (AR) and Virtual Reality (VR) Technologies

Immersive technologies like AR and VR are becoming increasingly prevalent in both consumer and enterprise settings. For instance, the Ray-Ban Meta smart glasses integrate AI assistants.

What to do now:

- *Use Meta Ray-Ban*: get hands-on experience with the technology and experience how it will change your work and personal life. Encourage teams to experiment with VR headsets to familiarize themselves with immersive experiences and identify potential applications within your business.

- *Assess Enterprise Use Cases*: explore how AR and VR can enhance training, collaboration, and customer engagement, considering industry-specific applications that could provide competitive advantages. Walmart, for example, uses Meta's Quest Pro headset to train new employees, reducing costs by a significant amount.

3. Redefining Customer Experiences

AMI enables hyper-personalized, real-time interactions, turning customer engagement into deeply tailored, value-driven experiences. We are moving from a world of one-to-many or one-to-few, to one-to-one experiences and content, enabled by AMI.

What to do now:

- *Get Your Data House in Order*: conduct a data audit to assess the quality, relevance, and usability of existing data. Break down data silos to create a unified view of the customer across all touchpoints and implement data governance frameworks to ensure compliance, security, and ethical usage.

- *Leverage AI for Personalization*: implement AI to analyze customer data and deliver contextualized, individualized recommendations across multiple touchpoints.

- *Invest in Omnichannel Integration*: ensure seamless communication between online and offline channels, so customers can transition effortlessly between environments (e.g., physical stores and digital interfaces).

- *Create Value Beyond Transactions*: shift from purely transactional approaches to building relationships by offering proactive, meaningful insights and services that demonstrate understanding and care for customer needs. Leverage these longer-term relationship building efforts to gather more zero- and first-party data that will be critical to giving you a competitive edge in the age of AMI.

4. Building Operational Agility

AMI enhances an organization's ability to respond dynamically to market conditions and internal demands.

What to do now:

- *Adopt Real-Time Analytics*: implement systems that gather and process data continuously, providing actionable insights into customer trends, supply chains, and resource allocation.

- *Embrace Resilient Design*: build flexibility into your business processes and infrastructure to adapt quickly to evolving technological and market landscapes.

5. Workforce Transformation

The workplace of the future will be fundamentally redefined by AMI, ushering in an era where every employee is empowered with personal AI assistants that follow them across devices, contexts, tasks, and even employers. These assistants will seamlessly integrate with organizational systems, creating a workplace where software and tools are dynamically customized to meet unique needs. This shift requires operational and support functions like HR and IT to adopt a future-forward approach, designing agile systems and policies that accommodate this unprecedented level of personalization and collaboration.

What to do now:

- *Prepare for Personal AI Assistants*: employees will arrive equipped with their own AI assistants, capable of handling everything from scheduling to drafting complex reports. Organizations must:

  - *Ensure Interoperability*: develop standards and protocols that allow personal AI assistants to integrate securely and efficiently with enterprise systems.

  - *Adapt Collaboration Tools*: equip employees with platforms that facilitate seamless interaction between personal and enterprise AI, fostering real-time cooperation.

  - *Balance Personalization with Security*: create policies that support the personalization benefits of these assistants while safeguarding sensitive company data.

- *Leverage AI for Rapid Software Customization*: the rise of generative AI will enable organizations to create custom software solutions at unprecedented speed. Organization must:

  - *Upskilling IT Teams in Context-Aware Development*: beyond basic code generation, IT teams will need to learn how to create software that interacts seamlessly with AMI systems. This involves incorporating advanced AI capabilities such as contextual understanding and dynamic response mechanisms into applications.

  - *Upskill IT Teams in Code Generation*: train teams to use generative AI tools for building and iterating on software tailored to organizational needs.

  - *Streamline Development Processes*: implement agile frameworks that allow rapid prototyping and deployment of AI-driven tools.

  - *Cross-Functional Collaboration for AMI-Ready Solutions*: while collaboration is common in current software development practices, designing for AMI requires engaging teams to think beyond departmental needs.

Software must support interconnected systems and diverse user contexts, making the co-creation process more complex and multi-disciplinary.

- *Create a Future-Ready Culture*: organizations must foster a culture that embraces the rapid pace of technological change:

    o *Champion Adaptability*: encourage employees to experiment with new AI tools and approaches, normalizing innovation as a daily practice.

    o *Empower Teams with Resources*: provide access to learning platforms, AI assistants, and collaborative tools that help employees thrive in an AMI-enabled environment.

    o *Emphasize Inclusivity*: ensure that AI systems and policies consider diverse perspectives and create equitable opportunities for all employees.

6. Addressing Ethical and Privacy Concerns

The reliance on personal and contextual data in AMI raises critical questions about transparency, security, and inclusivity.

What to do now:

- *Implement Privacy-First Practices*: design AI systems that prioritize user consent, data protection, and transparency from the ground up.

- *Create Ethical AI Standards*: develop frameworks to minimize bias, ensure accountability, and align AI systems with organizational values and societal expectations.

- *Engage Stakeholders Early*: involve customers, employees, and communities in discussions about how AI systems are designed and implemented, fostering trust and collaboration.

7. Reshaping Business Models

AMI opens new possibilities for revenue generation and value creation, requiring businesses to rethink how they deliver products and services.

What to do now:

- *Apply Design Thinking to Innovate New Models*: use a structured design thinking approach to identify unmet customer needs, ideate potential solutions, and prototype business models that leverage the unique capabilities of AMI.

- *Collaborate on Ecosystem Solutions*: engage with partners through co-design workshops to explore mutually beneficial AI-driven ecosystems, ensuring diverse perspectives and alignment on shared goals.

- *Focus on Iterative Testing and Refinement*: continuously prototype and test new value propositions, leveraging AI insights to refine offerings based on real-time feedback and changing customer behaviors.

8. Preparing for Personal Impact

AMI will influence how everyone works, learns, and lives, requiring us to adapt to new norms and expectations.

What to do now:

- *Adopt a Growth Mindset*: stay curious and open to learning about new tools, systems, and ways of interacting with technology.

- *Optimize Personal Productivity*: explore how AI-powered systems can streamline your routines, whether through smart assistants, productivity tools, or adaptive learning platforms.

- *Advocate for Inclusion*: as these technologies emerge, ensure they are accessible and beneficial to all, reducing digital divides and fostering equitable opportunities.

AMI represents more than just a technological shift—it's a reimagining of how humans interact with the world. This transformation integrates intelligence seamlessly into our environments, enabling hyper-personalized, scalable, and intuitive experiences. Whether through empowering individuals with proactive AI assistants, enhancing customer engagement with real-time insights, or reshaping industries with adaptable business models, AMI stands as a cornerstone of civilization's next big disruption. Embracing this evolution isn't optional; it's a necessity for staying relevant in an increasingly AI-driven future.

The path forward requires collaboration, innovation, and a shared commitment to building a better future. For individuals, this means cultivating a growth mindset and embracing opportunities to work with AI as a partner. For businesses, it means fostering ecosystems of trust, creativity, and resilience. As we collectively design the framework for this transformative era, we have the opportunity to shape AMI into a tool for empowerment—technology that enhances our humanity, enriches our experiences, and drives progress in a way that is as ethical as it is groundbreaking.

About the Author

Matt Collette is a seasoned leader in digital transformation and a trailblazer in AI solutions for marketing and communications. As the CEO of Sequencr AI, Matt leads a forward-thinking technology consultancy dedicated to empowering marketing and PR teams to unlock the full potential of artificial intelligence.

With a career spanning leadership roles in Asia and North America, Matt has consistently pushed the boundaries of innovation, earning accolades such as Asia's Digital Agency of the Year and Innovator of the Year in PRovoke's Asia-Pacific Innovator 25. Previously, he was head of digital for Edelman Canada and global head of digital growth, where he pioneered AI integration across the firm's global operations.

Website: www.sequencr.ai
LinkedIn: https://www.linkedin.com/in/mattcollette/

CHAPTER 7

THE OPPOSITE OF COOL: HOW AI TRAINING DATA CAN SHAPE (AND SAVE) THE FUTURE OF HUMAN-MACHINE TEAMING

By Dave Cook
Founder, The Training Data Project
Washington DC, USA

> *This is ourselves ... under pressure.*
> —Queen with David Bowie

Everyone has a story about how they got started in artificial intelligence (AI) and machine learning (ML). For some, their journey was inspired by the eerie presence of HAL 9000 in *2001: A Space Odyssey* or the quirky humor in *The Hitchhiker's Guide to the Galaxy*. Others were drawn in by the adrenaline-fueled worlds of *The Matrix* or *Blade Runner*. Most of these stories are connected to something fascinating—an interest in logic, math, machines, or code.

My story?

It's not that cool. I started with policing and crime data.

I didn't view what we were doing as AI/ML at the time, even though we used advanced computer models to analyze data and make predictions. Back then, people were not ready to embrace AI in policing, especially with movies like *Minority Report* shaping public perception. I led a team for a local law enforcement agency that prepared data for models to identify crime patterns and trends. We used geographic information systems (GIS) and mapping tools to support a police accountability system known as CompStat (short for "computer statistics"). Pioneered in New York, this methodology drew upon James Q. Wilson's and George Kelling's broken windows theory. The theory suggests that visible signs of disorder, like broken windows, indicate a lack of care in a neighborhood, which can lead to increased crime. The idea was that addressing small problems early on could help prevent larger issues in the future.

CompStat gained popularity among police departments across the country and was credited with reducing crime rates through data-driven strategies and focused interventions. However, its reliance on numerical data often came with drawbacks. The system arguably had a disproportionate impact on minorities and economically disadvantaged communities, reinforcing systemic inequities and fostering distrust. Many critics felt it prioritized meeting quotas over community engagement and thorough investigative work, raising concerns about fairness and the unintended consequences of data-driven systems.

Fast forward nearly 30 years, and I see parallels between "broken window" neighborhoods and today's AI landscape. AI has rapidly expanded, affecting almost every aspect of our lives. However, is everyone genuinely thriving in these AI "neighborhoods," or are we leaving people behind?

We now use AI and machine learning to make consequential decisions. Errors in AI and ML systems can influence creditworthiness or access to aid and insurance. The signs of neglect in our AI systems—our "broken windows"—are becoming more evident daily. Like a bustling freeway overpass looming over a declining community, technological brilliance and systemic flaws often coexist. We must ask ourselves: are these future AI neighborhoods places where we would want to live?

Fixing these challenges begins with the foundation of AI—its training data. For AI to truly serve society, its training data must be high quality, broad-based, and treated as a continuous commitment—including a commitment to fair treatment of the workforce that produces it. While thoughtful data curation may not be glamorous, it is essential for reducing AI risks. Poor-quality data sourced in an exploitative manner serves as a weak foundation. It creates flawed systems and exacerbates inequality. And in AI, bad data is often worse than no data.

This chapter examines how improving training data can help address the "broken windows" in AI. It highlights not *perfect,* but *better* practices that can mitigate potential vulnerabilities in training data development. The chapter emphasizes the importance of proactive measures to tackle these issues early on for better long-term human-machine teaming (HMT).

I also explore how high-quality training data can unlock AI's potential to create new opportunities in the US and global workforce. We can develop systems that reflect our values, build trust, and enable progress by addressing the vulnerabilities in training data. And that, as it turns out, *is* pretty cool after all.

Under Pressure: Aligning AI Systems with Training Data for Successful HMT

Training data begins with data labeling. Data labeling is both a human computational process and a machine learning-enabled process. Labeling creates training data to "teach" a machine what to do and how to behave. There is an art, a science, and a tension to good data labeling. A tension exists between what is labeled by machines versus humans and in producing data both quickly and accurately. While machines can label data much faster than humans, human labelers (i.e., humans-in-the-loop or HITL) add creativity, associative thinking, and judgment—the things that are essential to future HMT success. On top of this are tools, techniques, and workflows that ensure quality through code, statistical metrics, and human expertise. In short, it's messy. MIT research scientist Andrew McAfee said, "The world is one big data problem." *Training data* is no less messy. It demands constant attention.

This tension permeates every training data campaign or job, like a great rock song's beat or bass line—mechanical, intentional, and thoughtful. Think Queen and David Bowie in "Under Pressure." For instance, training an AI to identify aircraft on a tarmac requires precision in defense applications. Machines may take a first pass by labeling objects across a dataset, but HITL must refine and verify the results. This may slow the process, but this collaboration ensures the data's accuracy and creates a verified foundation for effective AI models. We must master this tension between HITL and machine-learning-enabled processes, letting machines do what they do well and letting humans do what only humans can do, if we want HMT to succeed and scale. The science of labeling requires a balance of machine efficiency and human guidance, paired with management science and robust processes to ensure high-quality training data.

The stakes in training data are high because its effects ripple through the entire AI pipeline. Poorly curated data doesn't just result in small mistakes—it propagates to become flaws that impact modeling, testing, evaluation, and deployment. According to Gartner research, poor quality can cost companies an average of $12.9 million annually. Poor-quality data impacts the AI's decision-making, creating significant financial loss. Eighty percent of the time spent on an AI project goes to cleaning and preparing data. That shows just how important training data is to future HMT. This is why training data alignment is critical. It ensures datasets are representative, reliable, and effective for their purposes. Engineers wouldn't build a bridge with structural weaknesses, so we shouldn't create AI systems without solid training data alignment.

The "Moneyball" Moment: Addressing Data Tension, Scarcity, and Sustainability

Any experienced AI/ML practitioner will tell you that, much like baseball, AI/ML is about getting on base. Home runs are thrilling—the sharp crack of the bat, the crowd holding its breath in anticipation, and the ball soaring past the outfielders. However, as the movie *Moneyball* illustrates, you don't win by simply aiming for home runs. Baseball is characterized by strategic trade-offs, balancing tensions, and leveraging limited resources to achieve success.

A key lesson from *Moneyball* is that we can find better solutions to complex problems by questioning conventional wisdom. Today, artificial intelligence is at a similar crossroads. There is a pressing need to think differently about efficiently and effectively accumulating more "runs" rather than merely pursuing "home runs." It's about making less spectacular yet more consistent and fundamental gains. In this context, focusing on AI training data is essential for consistently getting on base.

AI systems today face their own "Moneyball moment," trying to maintain data equilibrium—the push and pull of competing priorities—and managing data scarcity, where high-quality, real-world training data is increasingly limited. These challenges are compounded by growing energy and infrastructure demands, underscoring the urgent need for more sustainable, efficient, and innovative approaches to training data. Just as *Moneyball* revolutionized baseball by rethinking the fundamentals, the field of AI must now rethink how to address these pressing training data challenges and enable success.

Balancing the Scales: Managing Data Equilibrium for Smarter HMT

Data equilibrium highlights the challenge of balancing competing priorities in AI, such as privacy versus utility. Consider healthcare AI, where the stakes are critical: it must deliver precise diagnostics while adhering to strict privacy regulations. Recognizing and managing these demands requires more than technical expertise—it calls for strategic decision-making rooted in accountability and sound ethical judgment.

Data equilibrium is about using training data to elevate AI systems effectively. This involves improving quality, mitigating risks, and fostering continuous refinement while upholding accountability and ethical principles. It marries business and programmatic management with technical management. Success depends on curating data that is accurate, representative, and diverse. Addressing biases, incorporating edge (i.e., novel) use cases and situations to enhance adaptability, and ensuring privacy safeguards are critical to aligning AI systems with real-world needs.

Managing data equilibrium will create more intelligent, accountable AI systems for HMT. High-quality, well-curated data that reflects current conditions will ensure these systems meet operational requirements while maintaining ethical standards. This approach is essential for solving today's AI challenges and delivering solutions for complex, real-world demands.

Data Scarcity and Reasoning: The Looming Challenge and Opportunity for Better HMT

AI thrives on high-quality data, yet the supply of natural, reliable datasets is diminishing. By 2028, experts predict a critical shortage of high-quality text data for training models. Training LLMs also demands vast resources, consuming enough electricity to power a US home for 41 years and over 700,000 gallons of water for cooling servers. These figures underline the unsustainable nature of current practices and the need for more intelligent and efficient solutions.

One approach to addressing data scarcity is shifting focus toward smaller, curated, high-quality datasets. This would reduce resource demands. It aligns with research showing that AI trained on precise, diverse, and relevant data performs better. For example, a recent study found that reducing dataset size by 40% while improving its quality cut energy consumption by 30% without compromising model performance. Systems designed this way are less prone to hallucinations as they process fewer low-quality inputs. Similarly, a Microsoft study demonstrated that a pared-down AI system outperformed larger ones in reasoning tasks, reinforcing the value of smaller, focused datasets.

Another approach to scarcity is capturing logical thought and analysis when labeling data. This "labeling with reasoning" approach asks labelers to explain their thinking and provides a deeper contextual understanding for each label. A software engineer might be prompted to write a program that efficiently solves a complex logic problem. A historian might be asked to verify the reliability of sources that affect the credibility of labeled data. The answers—and, more importantly, how to reach them—are incorporated into the AI training material for things like LLMs.

Recognizing and managing data scarcity through improved quality, efficiency, and contextual reasoning offers a clear opportunity. By adopting smarter training data strategies, we can create systems that balance operational demands with sustainability, ultimately leading to more equitable and impactful AI solutions.

Elevating AI Performance: Metrics and Standards for Better Training Data

With clear quality measures, AI systems can avoid unnecessary risks. Managing risk iteratively through training data metrics is key to improving HMT. Training data must prioritize accuracy, consistency, and diversity. Mislabeling undermines AI performance, making quality-check and consensus tools for HITL labelers and subject-matter expert review vital for ensuring accuracy and resolving ambiguities.

Consistency drives more innovative solutions and better outcomes. Tools such as inter-annotator agreement metrics promote uniform labeling while dynamic review routing assigns tasks to skilled labelers, reducing errors and boosting efficiency. To achieve greater collaboration and transparency, standardized trust scores and interoperable platforms are essential, supported by regular audits to ensure data reliability.

Looking ahead, advanced quality metrics will empower AI systems to solve challenges more effectively. Focusing on completeness and diversity can reduce bias and improve AI's adaptability to real-world scenarios. Sensitivity analysis will identify and address common errors, lowering risks across applications. Coupled with adaptive systems for real-time feedback, these innovations will enable AI to support smarter workflows and enhance everyday life, ensuring a future where technology truly works for everyone.

Stopping Data Cascades: The Key to Strengthening HMT

Data cascades—compounding errors caused by poor-quality training data—present severe risks to AI systems. Studies show that 92% of AI practitioners encounter these issues, with 45% experiencing two or more in a single project. These cascades often lead to delays, budget overruns, and ethical challenges. For example, mislabeled medical data has skewed diagnostic models, creating risks that could endanger lives. Such failures often stem from undervaluing data quality and relying on flawed practices.

High-quality training data can be the most effective way to prevent data cascades. Practitioners can catch issues early by integrating rigorous validation protocols, adaptive controls, and iterative reviews. Ensuring diverse and representative datasets adds another layer of robustness, reducing the likelihood of errors snowballing across systems. These steps help manage risks across technical, financial, and ethical dimensions, creating stronger and more reliable AI.

Moneyball shows success lies in smart strategies and focusing on the fundamentals. This means rethinking how we should handle data quality and sustainability for AI. The broken windows theory provides a helpful analogy here: fixing minor issues now, like inconsistent labeling standards, can prevent larger failures. By emphasizing robust governance practices, we can create complex AI systems that align with operational goals.

Skilled Labor, Smarter AI: Prioritizing the Workforce That Can Make HMT Possible

Just as bands depend on skilled musicians and baseball teams rely on talented players, AI systems are built on the expertise of a HITL workforce. Unfortunately, data labeling is often regarded as unskilled "ghost" work. Just as businesses must certify how goods are sourced, we should know if the data that trains AI systems was curated responsibly. A robust training data supply chain can be established by treating data labeling and curation as skilled labor essential to effective human-machine collaboration.

Reframing data labeling as skilled labor is essential for enhancing training data quality. Overlooking data labeling and allowing it to live "in the shadows" creates persistent inequities that either now or will permeate our training data supply. These roles require precision, cultural understanding, and problem-solving skills backed by comprehensive training, fair compensation, and professional development. Recognizing the significance of this work would improve AI system quality while promoting economic mobility and generating more opportunities for underrepresented groups. Labelers are not statistical distributions or categories that flatten into numerically convenient measures of diversity. Labelers are a cognitive workforce that ensures AI systems can effectively address the complexities of real-world applications.

Objective truth in AI arises from the collective insight of a diverse workforce rather than from isolated viewpoints. The DC Mayor's Advisory Group on AI Values Alignment, which I am proud to be a member of, underscores the importance of accountability in AI systems, including the data that trains them. A skilled HITL workforce is vital to upholding these principles, fostering public trust, and ensuring that AI systems align with societal values. By tackling these AI-broken windows—inequities, undervaluation, and representation gaps—we can pave the way for an AI future that boosts human-machine collaboration and empowers individuals to thrive in AI training data careers.

Shaping the Future by Smartsizing Training Data for Human-Machine Collaboration

As AI advances rapidly, we must ensure that the "human" in HMT remains central. AI for the real-world hinges on solving one major challenge: training data. Whether in business or government, it's the Achilles heel. Everyone knows it's a problem, but solving it takes time, expense, and will. Our ability to live better and work smarter with AI relies on better training data curation—data labeling, quality controls, feedback, and incorporating edge use cases that cause AI models to learn and be more performant. AI will help us by better reflecting us from the start.

I recommend we tackle these challenges by focusing on creating high-quality, inclusive, and well-maintained training data, treated as a public good—something that benefits everyone and remains accessible within reason and without exclusion. Openness around training data will make AI systems fairer, more reliable, and more precise. The future of human-machine collaboration relies on the decisions we make today—it's ourselves—under pressure—to strengthen the fundamentals by prioritizing open, shareable training data curated by highly skilled labelers before small cracks lead to collapses.

Training data standards promoting interoperability can ensure consistent, trustworthy training data, while traceability mechanisms can link AI outputs to their original training data, fostering trust and accountability. Open-source tools can equally enable independent evaluation, promoting transparency and accessibility. By making small changes and investments now, we can fix AI's broken windows and build systems that foster HMT worth celebrating. And that wouldn't just be cool. It would change the course of our AI future at a time when it desperately needs it.

About the Author

Dave Cook is a technology expert with over 30 years of experience in advanced analytics and AI/ML. He supports classified programs for the Defense Department and Intelligence Community and teaches geospatial intelligence and AI/ML at the University of Maryland. In 2023, he founded the nonprofit Training Data Project to promote trust in AI/ML through education, standards, advocacy, and open-source tools. Dave advocates for diverse data labeling workforces, supporting fair treatment and inclusion.

Dave has tackled complex AI/ML and data challenges across government and the private sector throughout his career. An 18-time Marine Corps Marathon finisher, he is adamant that AI/ML is definitely more of a marathon than a sprint.

Dave serves on the DC Mayor's Advisory Board on AI Values Alignment and holds multiple advanced degrees and certifications in AI, data science, and geospatial intelligence. He lives in Georgetown (DC) with his wife, Noami, and their dog, Drake.

Email: dave@trainingdataproject.org
Website: www.trainingdataproject.org

CHAPTER 8

UNLOCKING YOUR BEST SELF: WELLNESS, EPIGENETICS, AND AI FOR SUSTAINABLE SUCCESS

By Jean-Michel JCA Davault, MBA
Engineer, Serial Entrepreneur & Business Leader
Brittany, France

> *Bouncing back after failure offers an opportunity to reinvent oneself and become a better version of oneself. Innovation in wellness and technology serves to create a sustainable and harmonious future for all.*
> —Copilot, October 25, 2024

What's a "Best Self" State?

More than ten years ago, I might have defined the "best self" state like this: my "best self" is the state in which I reach my full potential by aligning my values, personal and professional goals, and overall well-being.

I would have assumed my "best self" stemmed from an inner authenticity of being myself, my values, and my beliefs, without being influenced by external expectations. At that time, I am unsure if I would have spoken about my personal growth or development as I was pursuing my passion to develop collaborative software and provide for my family as a sense of accomplishment.

In 2012, I started to discover that this concept of "best self" also included "physical and mental well-being." At this time, I was working on a pre-AI project with black boxes (a constraint programing project). It was incredibly consuming, but even still, taking care of my physical and mental health was something I believed I was able to handle. After all, I thought I'd done a lot of work in these areas.

Alongside crisis management training at IBM, I was quite aware of the importance of balanced food, thanks to my mother who had taught me when I was six years old and frequently sick. I had thought I was already familiar with the importance of regular exercise and health practices. I had visited 49 countries and learned practices to master stress management with meditation, dô-in massage, yoga, and sport. From a traditional standpoint, I was well-educated with great (even amazing) teachers and professors (Prytanée de la Flèche, HEI Engineering school, MBA from EDHEC, professional certifications, many years MVP ...). Additionally, I was dedicated to continuous growth and constantly seeking to learn and develop myself through education, reading, and acquiring new skills.

As I was among the first to move from trusting the physical world to trusting software, from harnessing real-world skills to digital skills, from engaging in traditional management to functional and distributed management, I was aware of the importance of positive human relationships. When daily you work with people from all around the world—like my developer colleague Ashok Hingorani from India or Frédéric Sitbon in charge of East-European countries for an important collaborative software department at Microsoft—you understand you must cultivate healthy and rewarding relationships with others, based on respect, trust, and open communication.

When I co-wrote in 2009, "Microsoft Office Groove—Collaborative Trust in Networks," I had in mind human trust, digital trust, collaboration, people's skills, new ways to organize work, and perspectives of new sustainable business models. I intellectually understood what my contribution and impact could be. I also realized that humankind's talents and skills were dispersed inside each person.

To make a positive impact on society, whether through work, volunteering, or other forms of contribution, you need to be people-centric. Intellectually, this is something most of us already know. In practice some of us actually do it as well.

At Pre-AI Time, What Was My "Work-Life Balance"?

At the time of that pre-AI project, I was unsure if I had integrated a sufficient work-life balance. I was conscious that it was necessary. Did I really understand the warnings and signals I was receiving? I was unaware if I had found a harmonious balance between professional responsibilities and personal activities, a balance that would allow me to fully enjoy each aspect of life. In short, I might have thought that my "best self" was a version of myself where I was looking to feel fulfilled, aligned with my values, and able to contribute positively to my environment while taking care of my overall well-being. I probably thought I was operating in this zone.

And then the crash happened …

My work situation became unbearable due to a toxic colleague, forcing me to seek an escape from what felt like a forced employment contract. After enduring 15 months of this 'progressive hell' and lengthy negotiations, I finally secured a freelance contract, hoping for more autonomy and freedom. However, this arrangement was short-lived. Without warning, I was shifted into a traditional employment contract, complete with the hierarchical structure I had desperately tried to avoid. It feels like a bait-and-switch, leaving me feeling trapped and powerless. As this all happened ten years ago, I am now free to speak about it.

Ten days felt like ten years. I was trapped in a living nightmare, each day blurring into the next within the confines of my own personal coffin. The experience aged me profoundly, leaving me feeling like an old man, drained of life and vitality. I had increasingly severe muscle pain. My Thai massage sessions to strengthen energy circulation, to strengthen stress evacuation, to breathe better were reaching their limits. My hair was falling out in clumps. My face was becoming dull, and deep wrinkles were appearing. My sleep was fragmented and broken up by nightmares. My ability to concentrate was limited to about thirty minutes a day. I was overcome by enormous fatigue. Severe joint pain emerged.

The doctor did not really know what to say because technically, with his medical criteria, I was "still fine." He addressed the notion of professional overwork. He suggested solutions that could have secondary consequences for my body and even brain over the next ten years. He told me that I was undergoing a particularly violent bout of stress but that, as a doctor, the symptoms, although objective, observable, and visual, were outside the scope of a disease. The word "burnout" was mentioned. (I encourage you to read the definition of "burnout" from the World Health Organization.) Curious by nature, with a combative temperament, I tried to understand what was happening to me.

Introduction to the "Best Expression of Your Body" Epigenetics

Luck smiled at me because I was able to determine what was happening to me: intense oxidative physical stress which. impacts the expression of my genes ... epigenetics ... Hum, what is that? Let's continue, step by step.

One company proposed a "certified solution" to reduce my physical oxidation level in a healthy and natural way. This is how that solution worked: you measure your oxidative level with a machine that the developer won a Nobel Prize for creating. This machine is called a S3 BioPhotonic Scanner (not AI managed). The fourth generation of that scanner will be AI managed in 2025 and will be called "PRYSM iO". It is anticipated that at least 10 million units will be utilized over the next three years. An amazing step forward to measure the impact on our body of certified secure solutions.

You take a solution of a selected scientifically proven set of 31 nutriments during a minimum of three months (85% of the body cells are then renewed at least once). These nutrients help to strengthen nine essential classes of body functioning: antioxidant defenses, cognitive functions, the cardiovascular system, vision, metabolism, thyroid functioning, bone preservation, skin maintenance, and immune functions. At the end of the three months, you would have four evenly spaced measurements (one every three weeks).

Because I am impatient, I was "warned" that it would necessarily take three months. This is because the human body cannot technically run these cycles of normal cell renewal more quickly.

My measurements noticed a significant oxidative physical stress reduction over the three months. I left the first two red and orange zones to start to be in the yellow zone. (Of course, there is a scientifically recognized score called SCS to be more precise.)

I noticed my hair stopped falling out. I noticed a set of improvements with the feeling of an in-depth regeneration. I noticed a correlation with less mental fatigue.

A few months later I'd become more capable of analysis and reflection, and I understood I had just touched the essence of gene expression due to factors in the internal and external environments.

Imagine your genes as light switches. Each one can be turned on, off, or even dimmed to control its activity. This is called gene expression, and it plays a crucial role in our health and well-being. Scientists are now able to influence this process, essentially fine-tuning our genetic activity. While we have the potential to affect up to half of our genes, current technology allows us to influence only a small percentage. However, even these subtle adjustments can have a profound impact. Think of it like adjusting the settings on a complex sound system. Even small tweaks to a few dials can dramatically improve the overall sound quality. Similarly, influencing even a few genes can significantly benefit a person's health.

To understand this process better, scientists use «gene chips» to map and analyze gene expression patterns. This helps them identify which genes are active and how they interact, paving the way for targeted interventions that can optimize our health and well-being. It is essential to specify here that no gene is modified in any way; it's the opposite of creating a superman or a superwoman or becoming an augmented human from within with connected devices. We are here at the source of the inner "physical best self." Our inner "physical best self" has an influence on our "mental best self" and vice-versa.

To conclude this part of the chapter, let me clarify: my aim in telling you this personal story from my life as well as an overview of epigenetics is to challenge you to enlarge your vision of your "best self" state. What's your "Wellbeing4You"?

Human, Cultural, and Genetic Mix of Innovation

Let's start and correlate my birth on July 21, 1969, the date of the first known human step on the moon, with where I was adopted in Montreal, Canada. As I could notice, many years later at the births of my two boys (but not for my two girls) with their epicanthus, I have Inuit genes. Epicanthus corresponds to the expression of a child's slanted eyes at birth that disappears or greatly diminishes around age 7.

My adoptive mother, today 93 years old, still alive while I'm writing these lines, is an amazing woman. She was among those women that contributed to change women's position in society, like her mother who was the first to drive a car at the age of 17 near Antananarivo in Madagascar.

My mother, Monique Lesport Davault, a nurse, also born in Antananarivo, was one of those rare women of here time who gave birth to five-kilogram Bamilekes babies in Cameroon, alone, in the middle of the African bush. She was the type of woman to decide at the age of 37 to adopt a child with her husband, Alain, my father.

My academic background is engineering and business. In my early career I was a top evangelist of collaborative software at Lotus, IBM, and Openwave). I was a co-founder of Hommes & Process. I pioneered advancements in collaborative web and cyber defense sectors. All of this helped me to be able to anticipate the arrival of AI and the importance of personalized wellness and well-being tools and solutions.

Transition to Wellness Based on Epigenetics and AI

As you may now guess, the combination of epigenetics and AI holds immense potential for revolutionizing wellness as developing solutions at the source of aging, for instance. I gave you a personal example where epigenetics allowed the understanding of how environmental factors and lifestyle choices can influence gene expression without altering the underlying DNA sequence. This means that we have some control over our genetic destiny through our daily habits.

By understanding epigenetics, either solutions providers or you can identify personalized strategies to optimize your health and well-being. AI algorithms can analyze vast amounts of data, including genetic information, lifestyle factors, and environmental exposures. This enables AI to identify patterns and correlations to imagine new solutions as personalized as possible.

AI Contributes to Producing Collagen Amino Acids from Plants

Let me provide you with an example. The collagen contained in the bio-adaptive skin care serum "Nutricentials Vitamin C + Collagen Pumps" is composed of fragmented collagen of plant origin. "Fragmentation" means that the collagen molecules have been reduced to a size that allows the skin to absorb it more easily to benefit from the collagen's amino acids. Plants do not produce collagen. But thanks to exceptional (AI-based) biotechnological (research) techniques, it has been made possible to make these plants produce the amino acids found in collagen from a synthetic fragment of human type 1 collagen.

Furthermore "bioadaptive" helps you to visualize the mechanisms that use plants to adapt to their environment and hence your skin (based on epigenetic long-run measures). Here are some bioadaptive plants: resurrection plant extract (Selaginella extract Lepidophylla), ginseng Root Siberia (Eleutherococcus Senticosus root extract), Rhodiola Rosea Extract (Rhodiola Rosea extract Rosea), Maral Root Extract (root extract of Rhaponticum Carthamoides), from the Extract of Chaga mushroom (Inonotus Obliquus extract).

The Power of Personalization Based on AI

AI-powered software can, for instance, provide personalized recommendations for diet, exercise, and stress management. AI-powered devices help to deliver internet of things (iOT) personalized solutions like massages or circulation of lymphatic fluids without pain (look at the ageLOC RenuSpa iO on the US market (https://nskn.co/ETtBgk) or the ageLOC WellSpa iO on the European market (https://nskn.co/wfv2Nv).

By integrating epigenetics and AI, highly personalized wellness plans, tools, or devices that target the root causes of health issues can be developed. I won't speak about AI-powered platforms that can provide continuous monitoring and feedback, helping individuals stay on track with their wellness goals because with such platforms people might lose their freedom or become addicts or even slaves. I prefer to concentrate on AI-powered solutions with more targeted scopes, for instance, to check a person's overall skin hydration, the impact of the sun on unprotected skin, or the analysis of a set of other day-to-day good health skin parameters. I invite you to test the VERA application from your iOS/Android store developed by Nu Skin Enterprise (see About the Author).

At AI Time, What Is My "Work-Life Balance"?

It's been ten years since I experienced that difficult burnout and epigenetic-backed resurgence. I invite you to make your own opinion about the shift in my work-life balance since that time. I invite you to compare and follow what I've published and how I've shifted to a new work-life balance and "best self" by comparing my professional Linkedin profile and my Instagram account or other publications over time.

On my Instagram account @wellbeing_for_you, you will find more than 3,100 publications, photos, and reels from the beginning of my restart to today. Expect to enjoy photos with a French background and see me living out a well-balanced entrepreneurial lifestyle. Today I am committed to propagating a well-balanced lifestyle for entrepreneurs.

In conclusion, it is critical that each of us put wellness at the center of our lives and the communities around us. It is essential we continue to have faith in AI and technological innovations as forces of good, as light will always survive through adversity. I encourage you to remain committed to your success and well-being. It is possible, especially when harnessing the power of AI with epigenetics.

About the Author

Jean-Michel Davault is a French entrepreneur and wellness advocate, known for integrating epigenetics and AI to promote sustainable success. As the founder of W4P4 (WellBeing4You at the Power 4), he leverages advanced technologies to enhance well-being and aging. With an MBA from EDHEC, he has pioneered advancements in the collaborative web and cyber defense sectors. Over the past decade, he has partnered with Nu Skin Enterprises, serving as an ambassador for their Nourish the Children initiative. His focus is on holistic health, aligning personal goals with team projects. Through Les Rebondisseurs Français, he promotes "Rebond," encouraging entrepreneurs to view failure as an opportunity to restart and improve. He also collaborates with the international Indian community to promote business opportunities in the 2025 Indian market opening with Nu Skin.

To test the VERA application from your iOS/Android store developed by Nu Skin Enterprise, for free, you can create an account based on this code FR3346890. My name will appear as a control parameter to you.

Email: jmdavault@w4p4.com
Website: www.w4p4.com

CHAPTER 9

REVOLUTIONIZING BUSINESS: THE POWER OF AI IN THE MODERN WORLD

By Malcom W. Devoe, PhD
Innovator, Educator, Wellness Visionary
Atlanta, Georgia

> *Some people call this artificial intelligence, but the reality is this technology will enhance us. So instead of artificial intelligence, I think we'll augment our intelligence.*
> —Ginni Rometty, former CEO of IBM

Over the last two decades of the 21st century, we have entered a shift from technology being a dedicated tool designed only to maximize efficiency to technology being the fundamental aspects that support day-to-day global business operations. Today, digital platforms are crucial for companies to access customers, manage supply chains, and analyze data. Furthermore, the advent of cloud computing, big data, and mobile technologies has driven businesses to expedite digital transformation. As innovation accelerates rapidly, companies have to remain nimble and agile, harnessing the latest technologies to remain the fittest in crowded markets.

One of the most important developments of technology in the last few years is the emergence of artificial intelligence (AI). According to Doug Rose, AI is a machine's ability to emulate characteristics of human behavior (2021). Many of us have been exposed to various AI devices and tools. We use these tools unknowingly in our everyday lives. Some examples of these include smart assistants such as Siri, Alexa, or Google Assistant. AI is also featured in an individual's personalized recommendations with TV, music, and shopping apps. When a person's TV app continuously provides them with suggested movies, music, or things to buy, then that means that the individual has an AI algorithm behind the scenes suggesting what to watch, listen to, or even suggest what to buy, which it learns through an individual's purchasing and browsing history. Other examples of AI tools that are used in an individual's everyday life are social media algorithms, navigation maps, email filters and smart replies, health and fitness apps, virtual customer support, security and fraud detection, and many others.

Not only has AI affected our everyday lives, but it has also influenced the way we do business. AI tools can change the way business owners do business by providing tools to automate repetitive tasks, improve decision-making, and help provide personalized customer interactions. Machine learning (ML) algorithms are used to analyze extensive datasets to forecast trends, and natural language processing tools may simplify customer service with the help of chatbots and virtual assistants. Thus, using AI-based analytics platforms, business executives can gain insights into trends in the market, customers, and operational efficiency, helping them make business decisions. As AI has become an essential factor in production promotion, it is destined to innovate in unthinkable ways.

Common AI Myths and Misconceptions You Should Know About

Many myths and misconceptions about artificial intelligence come from a combination of media portrayals, misunderstandings about what AI can and cannot do, and concerns about its potential effects. A common misconception is that artificial intelligence will cause mass unemployment by taking over human jobs. Although AI automates tasks, research shows that it enhances human work more often than it truly replaces jobs; it transforms jobs toward creative and problem-solving tasks (Bessen, 2019). As AI matures, so do the fields that study it, such as data analysis, machine learning, and AI ethics (Manyika et al., 2017).

Another misconception is that artificial intelligence is not much less intelligent than a human and has the ability to reason entirely independently. The truth is that most AI systems (machine learning models) are highly narrow and generally unable to generalize or transfer to other tasks than what they are trained to learn. AI systems, rather, can only work within a particular range of limits while human types of intelligence can explore beyond those limits (Russell & Norvig, 2021).

Another misconception is that AI systems are non-partisan. AI models, in reality, can mirror the biases that exist in their training data, frequently in a way that unintentionally strengthens stereotypes (O'Neil, 2016). This is paramount to making artificial intelligence tools as equitable and trustworthy as possible (Friedman & Nissenbaum, 1996), keeping in mind that no algorithm is perfect and every model has a certain degree of bias, which needs to be actively managed in the modeling process.

Finally, there are those who think that artificial general intelligence will outpace human control, maturing into a super intelligent system and presenting an existential factor on a global scale. However, the advanced AI development they envision is extremely speculative, and many agree that we are considerably far from developing AI with broad intelligence or agency (Deloitte, 2023).

It is these types of misunderstandings that demonstrate the necessity of public education and informed discourse to explain the capabilities—and, more importantly, the limitations—of AI. Awareness of what AI can and cannot do will better position society to reap the rewards of AI, as we spend less worrisome thoughts on its harms (Bostrom, N. (2014).

Benefits of AI for Businesses

Businesses have been presented with a unique chance through artificial intelligence to save costs and provide more efficient operational processes. For example, AI can automate many of the mundane and repetitive tasks associated with data entry, inventory management, customer support, and a variety of other day-to-day operations, allowing businesses to continue producing at a high level, with fewer resources (Marr, 2020). For instance, deploying chatbots that utilize AI to answer routine customer questions saves businesses the cost of recruiting more customer service personnel and also enhances response times. Furthermore, AI automation in accounting and marketing can cut down overhead, and when fully deployed, driving down costs, resources can be deployed from a business to other crucial areas such as product or customer acquisition (Huang & Rust, 2018).

Thus, businesses can significantly benefit from AI, as it provides them a way to know how their customers behave, so they can customize their offerings. Therefore, customer satisfaction improves. By correlations of consumer preferences, purchasing trends, and feedback results, businesses can personalize marketing techniques and improve inventory management with AI-powered analytics tools (Davenport & Ronanki, 2018). This will allow businesses to make better decisions that can have a direct bearing on revenue growth and customer loyalty. Additionally, AI simplifies the life of businesses through (the development of) advanced tools with which they can predict demand, optimize pricing, target potential customers, and ultimately enhance their competition against larger companies (Brynjolfsson & McAfee, 2017).

AI in Customer Service and Support

Businesses looking to improve customer engagement and support can adopt tools such as chatbots and virtual assistants powered by artificial intelligence. Natural language processing (NLP) works in conjunction with platforms like Intercom and Drift (Chaidrata et. al., 2022) to provide instantaneous responses to customer inquiries and take the reins for basic and repetitive interactions, including FAQs, product information, and order tracking. Because businesses can respond to these inquiries without human intervention, chatbots enable 24/7 support with much faster response times. Tools like these work especially well for businesses that have fewer customer service personnel, as they eliminate the need for dedicated support agents and maintain customer satisfaction (Ekemezie et. al., 2024). In addition, AI chatbots and virtual assistants assist businesses with customizing customer interactions based on data about the user behavior, preferences, and prior interactions.

AI for Marketing and Sales

With tools such as HubSpot and Mailchimp, AI is already changing the marketing and sales landscape, giving businesses an opportunity to implement campaigns on target and ensuring a well-personalized delivery. These platforms use machine learning algorithms to sift through customer data, including things like browsing history, purchase patterns, and engagement behavior, in order to design personalized marketing plans. Using this data, Mailchimp can automatically segment audiences and email campaigns to send specific emails that will most likely capture customers' attention (Marr, 2020). Using advanced predictive lead scoring and content recommendations, HubSpot also emphasizes creating deeper connections with customers using AI capabilities to personalize interactions in similar ways. This method of personalization enhances customer engagement and conversion pages, which makes marketing efforts more efficient (Patil, 2024). Moreover, AI's data-based insight gives businesses a greater understanding of their audience and helps them make optimum decisions.

AI in Data Analytics and Decision-Making

Data analytics tools powered by AI are changing the way businesses analyze their customers, sales trends, and financial performance. AI algorithms drive software solutions, such as Google Analytics and Tableau, to help companies better understand large sets of data ranging from website traffic trends to purchasing behaviors. A good example would be Google Analytics, which helps business owners by providing detailed information on customer interaction within a website, such as the amount of time spent on each page and the number of pages visited (Chaffey & Ellis-Chadwick, 2019). In addition to this, Tableau provides users with the ability to create a visualization that reflects real-time data, thus enabling businesses to quickly identify new trends and analyze how sales are at any given moment (DataRoot Labs, 2021). Being capable of processing and analyzing data rapidly and accurately helps businesses to respond better to customer requirements and market demands.

One of the powerful aspects of AI is machine learning; such predictive analytics is an integral part of business decision-making. Predictive analytics, for instance, can analyze historical data to predict future sales, seasonality trends, and even the likelihood of customers leaving (Sharda et al., 2020). AI tools leverage sales data to show businesses forecasts on when they can expect peak sales periods in AI tools. Such data, still from AI tools, benefits companies by allowing them to pre-adjust stock and staff accordingly. This is particularly helpful for small businesses, which often have fewer resources and need to be strategic about how they use them. These predictions will help business owners make informed decisions, which will lead to increased profitability and decreased wastage (Delen & Ram, 2018). With this type of data insight, businesses are now able to make strategic decisions with certainty.

AI for Process Automation

Businesses are increasingly utilizing AI process automation tools to automate daily tasks such as invoices, payroll, scheduling, and more. Secondly, many businesses use tools like Zapier and QuickBooks to automate repetitive processes rather than spending hours manually checking every transaction. A great example of this is QuickBooks, which is an AI-powered software that processes expense tracking, invoice creation, payroll processing, and other tasks that would take business owners hours away from effective operating (QuickBooks, 2021). Zapier is another one of the many tools similar to Integromat that allows different apps to communicate and allows businesses to migrate data from one platform to another without the need to manually enter information (Martinez, 2020). These are capable of autonomous management of simple and repetitive tasks, allowing businesses to remain accurate and allowing business owners to be free from administrative work.

When it comes to the timesaving and cost-cutting advantages of automation, business owners (who often wear many hats) are at an incredibly great advantage. AI process automation streamlines a business owner's processes efficiently, from sending invoices and scheduling to freeing business owners from mundane administrative tasks to focus on strategic growth activities. Automated task execution also lessens the risk of human errors that can be expensive and time-consuming to fix. Automation in payroll processing, for example, helps ensure that there are no mistakes due to manual errors while calculating salaries and tax, thereby reducing the risk of fines (Davenport & Kirby, 2016). With AI-powered tools working 24/7, businesses can speed up the processing of tasks, thus enhancing the turnaround time for a few critical processes. Thus, AI for process automation enables businesses to function with the efficiency and accuracy of larger businesses, which levels the playing field and supports sustainable growth.

AI for Inventory and Supply Chain Management

New approaches for AI are reinventing inventory and supply chain management by enabling businesses to optimize on-hand stock, reduce waste, and discover precise components to accurately predict demand. Applications such as TradeGecko and ClearSpider use historical sales data, seasonal changes, and market variations to develop machine learning algorithms on the future need of inventory. AI can help businesses even out their stock levels to prevent overstocking, which contributes to waste, and prevent stockouts, which ties up working capital in unsold inventory (HashStudioz, 2024). AI-powered demand forecasting also enables businesses to prepare for peak and slow seasons, so they can make smarter purchase decisions and manage their stocks effectively (EVIZI, 2024).

Besides smoothing inventory, AI tools often facilitate the supply chain and logistics. For example, ClearSpider provides visibility of where inventory is located, making tracking and fulfillment easier than ever. AI-enabled logistics support can help businesses automate the process of reordering and even facilitate direct integration between suppliers, which allows for faster communication and response times to certain levels (Ganesan, 2020). This enhanced efficiency reduces unnecessary delays, cuts down on logistical mistakes, and allows products to pass seamlessly between suppliers and customers. However, for some businesses, many of which may not have a dedicated supply chain team, these tools provide a consistent and effective means to manage and simplify complex logistics operations.

Businesses with limited budgets can use AI-enabled inventory management to pay for more items, give customers more of what they want, and have the bandwidth to grow (Marr, 2020). In the end, using AI in inventory and supply chain management helps small businesses to easily compete with the bigger firms by streamlining their workforce and maintaining more reliability.

AI for Financial Management

Artificial intelligence financial management tools are revolutionizing the accounting and bookkeeping space for businesses, by automating key functions such as expense tracking, invoicing, and reporting. AI is utilized to expedite these processes in Xero, FreshBooks, and many other platforms where expenses are automatically categorized, invoices are generated when products or services are sold or utilized, and financial reports are produced from real-time figures (QuickBooks, 2021). Xero links directly to business bank accounts and credit cards to automatically record expenses and categorize transactions. FreshWorks has similar automation for invoicing and payments, so businesses can get paid faster and use less manual data entry. This not only saves time for the business owner but allows them more time to strategize and run a growing business instead of dealing with day-to-day numbers (Hyland, (n.d.).

These AI-based technologies in financial management slash the time taken for fraud detection and forecasting, assisting small businesses. ML algorithms enable business owners to keep an eye on different transactions for any irregular patterns or exceptions that could appear, allowing the detection of fraud instances at early stages. As a result, it enables early prevention of fraudulent activity. Some AI tools, for example, will review transaction histories and flag any activity that appears to be out of character for additional fraud detection for a business without an entire team dedicated to preventing fraud (GeeksforGeeks, 2024).

Similarly, AI-powered forecasting tools analyze historical data and market trends to project future financial performance, equipping businesses to better budget, allocate resources, and manage cash flow (PDF. ai, 2025). Being able to predict money coming in and out of a business is good for all types of businesses, but it is particularly useful for small ones who are anticipating having little cash on hand. This approach to finance finally assists businesses in increasing profitability, lessening risk exposure, and making better use of scarce resources. (Marr, 2020)

Getting Started with AI

We all know that getting started with AI is intimidating for small business owners, but selecting the right tools can go a long way in improving operations and accelerating growth. Budget, ease of use, and scalability should be the guiding principles when selecting AI tools. Small businesses normally operate on tight budgets, making it imperative that the business owner prioritize tools with affordable subscription models, free trials, or pay-as-you-go options (US Chamber of Commerce, 2024). Simplicity is equally vital as user-friendly interfaces along with accessible customer support to expedite the learning curve and allow teams to implement the tool without deep-dive training. Scalability is yet another crucial aspect; selecting AI tools that can help grow a business will ensure the owner does not outgrow their instruments when the operations start scaling (Marr, 2020).

To effectively use AI, integration with existing software is essential. If AI tools combine naturally with existing systems such as customer relationship management (CRM) software, accounting platforms, or email marketing tools, it can lead to seamless operations by keeping manual transfer of data to the minimum and creating automatic real-time data synchronization (Davenport & Kirby, 2016). For instance, AI tools that integrate with QuickBooks for accounting or Shopify for e-commerce enable small businesses to provide improved financial management and marketing without needing to purchase entirely new systems. Compatibility can also help reduce setup costs in the initial stage and facilitate ongoing maintenance.

Various cheap and easy AI tools needed for businesses are available. Marketing and CRM tools, such as HubSpot, automate repetitive marketing tasks and provide core services to the sales team; accounting automation tools like Xero automates basic accounting processes while enabling accountants to offer value-added services; customer support applications, such as Zoho, address customer needs by providing insights from data (Forbes Advisor, 2024). For example, a company like HubSpot has capabilities powered by AI, such as personalized email marketing and predictive lead scoring, while Xero simplifies expense tracking and invoicing. Using these tools of AI technology, business owners can analyze how to use data in the best possible way to reduce operational costs, be instantly aware with data, and subsequently improve customer service and productivity.

References

Bostrom, N. (2014). *Superintelligence: Paths, dangers, strategies.* Oxford University Press.

Brynjolfsson, E., & McAfee, A. (2017). The business of artificial intelligence: What it can—and cannot—do for your organization. *Harvard Business Review.*

Chaffey, D., & Ellis-Chadwick, F. (2019). *Digital marketing: Strategy, implementation, and practice.* Pearson.

Chaffey, D., & Smith, P. R. (2020). *Digital marketing excellence: Planning, optimizing and integrating online marketing.* Routledge.

Chaidrata A. et. al. (2022) Intent Matching based Customer Service Chatbot with Natural Language Understanding. arXiv.

Chopra, S., & Meindl, P. (2020). *Supply chain management: Strategy, planning, and operation.* Pearson.

DataRoot Labs. (2021). *Data visualization with Tableau.* Datarootlabs.com.

Davenport, T. H., & Kirby, J. (2016). *Only humans need apply: Winners and losers in the age of smart machines.* HarperBusiness.

Davenport, T. H., & Ronanki, R. (2018). Artificial intelligence for the real world. *Harvard Business Review, 96*(1), 108-116.

Delen, D., & Ram, S. (2018). Predictive analytics in business: Theory and applications. *International Journal of Business Analytics, 5*(1), 45-60.

Deloitte (2023). Trustworthy Artificial Intelligence: Ensuring Fairness and Trust, www2.deloitte.com.

Ekemezie, O. et. al. (2024). *AI Chatbot Integration in SME Marketing Platforms.* International Journal of Digital Marketing

EVIZI. (2024). *Enhancing customer engagement: The role of AI and data in personalized digital marketing.* www.evizi.com.

Forbes Advisor. (2024). *Best CRM Software of 2025.* Forbes. www.forbes.com

Ganesan, K. (2020). *The business case for AI.* Opinosis Analytics Publishing.

GeeksforGeeks. (2024). *10 Best AI Tools for Fraud Detection in 2024.* www.geeksforgeeks.org.

HashStudioz. (2023, 2024) *AI in inventory management.* www.hashstudioz.com

Huang, M., & Rust, R. T. (2018). Artificial intelligence in service. *Journal of Service Research, 21*(2), 155-172.

Marr, B. (2020). *Artificial intelligence in practice: How 50 successful companies used AI and machine learning to solve problems.* Wiley.

Martinez, K. (2020). *What is Zapier? Automation Basics.* zapier.com

O'Neil, C. (2016) *Weapons of math destruction: How big data increases inequality and threatens democracy.* Crown Publishing Group.

Patil, D. (2024). *Artificial intelligence for personalized marketing and consumer behavior analysis*. SSRN Electronic Journal.

PDF.ai. (2025). *The Top 10 Free AI Tools for Financial Analysis in 2025*. www.pdf.ai

QuickBooks. (2021). Automation in accounting: Saving time and improving accuracy for small businesses. *QuickBooks*. quickbooks.intuit.com.

Rose, D. (2021). *Artificial Intelligence for business* (2nd ed.). FT Press.

Russel, S., & Norvig, P. (2021) Artificial Intelligence: *A modern approach* (4th ed.). Pearson.

Sharda, R., Delen, D., & Turban, E. (2020). *Business intelligence, analytics, and data science: A managerial perspective*. Pearson.

U.S. Chamber of Commerce. (2024). *Top Subscription Models Driving Small Business Growth*. www.uschamber.com.

About the Author

Dr. Malcom Devoe is a dynamic leader, educator, and innovator blending STEM education with mental wellness. As CEO of Devoe Digital Learning, he equips teachers and inspires students to master coding while his work with Military Consulting Solutions pioneers global digital workforce platforms to revolutionize education, healthcare, and job development. A faculty member at Morehouse College, Dr. Devoe combines his expertise in computational mathematics and digital game-based learning (DGBL) to create engaging tools that transform how students learn math and computer science.

Passionate about mental health, Dr. Devoe co-hosts the lively podcast *R U Serious?* with Dr. Daniel Upchurch, tackling mental wellness with humor and practical insights. Beyond academics, he's a hip-hop cardio instructor, blending fitness and fun to empower others. With a commitment to innovation and positivity, Dr. Devoe helps individuals shift mindsets, embrace well-being, and find joy in learning and living.

Email: devoedigitallearning@gmail.com
Website: https://www.devoedigitallearning.com/

CHAPTER 10

THE LAST INVENTION: AGI AND THE FUTURE OF HUMAN POTENTIAL

By Hassen Dhrif, PhD
AI Applied Scientist and Engineering Leader
Kirkland, Washington

> *The greatest danger in times of turbulence is not the turbulence; it is to act with yesterday's logic.*
> —Peter Drucker

Introduction: A Turning Point in Human Progress

Imagine a future where machines don't just assist us—they think and reason like humans. They learn, adapt, and even create in ways that enhance our lives and solve the challenges we face. That future is not as distant as it seems. It's one that's being shaped by artificial general intelligence (AGI)—the next major leap in artificial intelligence that promises to forever change how we work and live.

In this chapter, we will explore why AGI may very well be the last major invention humanity needs. As AGI evolves, it will not just augment our abilities; it could revolutionize the way we approach the world's most pressing challenges. AGI's potential is vast, but so are its risks. For professionals, business leaders, and citizens alike, it's essential to understand how we can harness AGI responsibly and effectively. The question isn't just whether AGI will change our future—it's how we will adapt and thrive alongside it.

What Is AGI and Why Does It Matter?

AGI is fundamentally different from both the artificial intelligence (AI) we encounter today and the large language models (LLMs) that power current AI assistants. While LLMs can process and generate human-like text, they lack true reasoning and general problem-solving capabilities. While narrow AI systems like voice assistants and recommendation algorithms are highly specialized, AGI is capable of performing any intellectual task that a human can do. Think of it as a supercharged brain—a machine that can reason, plan, learn, and even understand emotions.

Unlike today's AI, which excels at specific tasks but fails when faced with unfamiliar situations, AGI will be adaptable. It will be capable of learning from experience, thinking critically, and even creating new ideas. This makes AGI incredibly valuable in a world where knowledge and innovation are constantly evolving.

But why does this matter for you? Because AGI isn't just another tool—it's a partner that can enhance your potential, whether you're a professional looking to increase productivity, an entrepreneur driving innovation, or a policymaker navigating new ethical waters. AGI represents the next frontier of human progress—and while significant progress is being made, the timeline for achieving true AGI remains a subject of debate among experts.

How AGI Will Change How We Work

One of the most immediate ways AGI will impact society is in the workplace. For many professionals, AGI will work alongside them to take on repetitive or data-heavy tasks, allowing them to focus on the work that requires creativity, strategy, and emotional intelligence. Here's how AGI will make us work smarter:

Unlocking Human Creativity and Innovation
AGI will automate mundane tasks like data entry, customer service, and scheduling. In doing so, it will free up professionals to focus on creative, high-value activities that drive innovation. In industries like design, marketing, or product development, AGI can be a collaborator that suggests ideas, tests prototypes, and even identifies new market trends.

Enhancing Decision-Making
AGI can help professionals make better data-driven decisions. Whether you're in finance, healthcare, or logistics, AGI can analyze vast amounts of data in seconds, identify patterns, and provide insights that might take human analysts weeks or months to uncover. This leads to faster, more accurate decision-making that can transform entire industries.

Tailored Experiences for Entrepreneurs and Business Leaders
AGI can revolutionize how businesses operate. For entrepreneurs and CEOs, AGI could act as a personal advisor—optimizing workflows, predicting market trends, and even suggesting strategic partnerships. For small businesses, AGI can level the playing field by providing sophisticated tools once only available to large corporations, helping them stay competitive and grow faster.

A Personal Assistant for Work-Life Balance
As more of us juggle work and personal lives, an AGI-powered assistant can help us navigate daily tasks, streamline communication, and prioritize our time. By automating scheduling, handling emails, and managing projects, AGI will help individuals and teams achieve a healthier balance between work and personal life, making it easier to stay productive without burning out.

The Risks and Challenges of AGI

Despite the tremendous potential of AGI, we must approach its development with caution. AGI's very intelligence presents risks if it is not aligned with human goals and values.

The Alignment Problem
The central challenge of AGI is ensuring that it aligns with human values. An AGI system could learn to optimize for efficiency, but what if its defi-

nition of efficiency conflicts with human well-being? How do we ensure AGI's goals don't lead to unintended harm? This is where ethical governance comes into play—designing AGI to be transparent, accountable, and aligned with human interests. Leading AI labs are actively working on alignment through various approaches, including constitutional AI, reward modeling, and inverse reinforcement learning. However, a key challenge lies in ensuring that AGI systems maintain their alignment with human values even as they become more capable and potentially able to modify their own code or goals. This "alignment stability" problem remains one of the field's central challenges.

Job Displacement and Economic Disruption

Many fear that AGI will take away jobs, particularly in areas like manufacturing, customer service, and even professional services. While historical technological advances have typically created new jobs to replace those they eliminated, AGI-driven automation may be fundamentally different. Unlike previous technologies that automated specific tasks, AGI could potentially match or exceed human capabilities across almost all cognitive domains. This unprecedented scope of automation may require new economic models, such as universal basic income or other innovative approaches to ensure economic stability and fairness. AGI could enhance productivity, but we'll need to invest in reskilling and upskilling workers to ensure they're ready for the jobs of the future.

Ethical and Legal Dilemmas

As AGI becomes more capable, we will face questions about its rights and responsibilities. Can an AGI system be held accountable for its actions? Should it have legal standing? These are tough questions, and answering them will require careful thought, open dialogue, and robust frameworks for governance.

Leveraging AGI to Solve Global Challenges

AGI has the potential to address some of the most pressing challenges humanity faces. From combating climate change to eradicating poverty, AGI can process vast amounts of data and develop innovative solutions to problems that have long evaded human efforts.

For instance, in healthcare, AGI could revolutionize drug discovery, leading to faster treatments for diseases. In environmental sustainability,

AGI could optimize resource usage, reduce waste, and develop new technologies for clean energy.

By harnessing AGI, we can move towards a more sustainable and equitable world. The question is: how can we guide AGI's development to ensure these benefits are realized by all?

Conclusion: AGI and the Path Forward

AGI promises to be humanity's last major invention, but how we develop and integrate it will determine its impact on the world. If we approach AGI responsibly, with a focus on ethical development, human collaboration, and sustainability, we can ensure that it enhances our lives and helps us solve the challenges of the 21st century.

The future of work, innovation, and human potential is tied to how we adapt to AGI. It's not about whether we should embrace it—but how we can work smarter, live better, and shape a future where humans and AGI thrive together.

About the Author

Dr. Hassen Dhrif is an AI leader and generative AI expert with extensive experience in software engineering, data analytics, and machine learning. As a senior applied scientist at Amazon, he drives innovations in Alexa's large language models and multimodal AI applications. Prior to Amazon, he served as the head of AI at Writer, where he successfully integrated large language models into commercial products. His diverse experience includes leading a groundbreaking multimodal AI project in cardiology at Weill Cornell Medicine, demonstrating the practical applications of AI in healthcare. Holding both a PhD in computer science and an executive MBA, Dr. Dhrif combines technical expertise with business acumen to develop AI solutions that address real-world challenges and drive organizational innovation.

Email: hassen.dhrif@gmail.com
LinkedIn: https://www.linkedin.com/in/hassendhrif/

CHAPTER 11

DRIVING DIGITAL TRANSFORMATION: A VISIONARY IMPACT ON GLOBAL FINTECH, BANKING, AND AI-POWERED INNOVATION

By Parham Emami, MMSci, BEng-CE, PMP
Nobel Tech Founder, Lecturer, AI Innovation Consultant
Toronto, Ontario, Canada

Technology and Me: Essential Lessons Learned

I still remember when I was tasked with leading the build and delivery of the world's first cloud-based secure mobile payment technology and product about 20 years ago. It had implications worth billions of dollars for the business and industry. At that time, I was a novice project and product manager.

Since then, I have led the development of many other new products and services, often building the next or sometimes even two generations ahead in technology, always aiming for a positive impact. The reason is that technology adoption and refinement take time. As an innovator, whether an individual or an enterprise, you must constantly experiment with new technological possibilities and be ready for the next wave. That is the key rule for proactively staying relevant.

Like any other first wave of technological breakthroughs, I had to face many unknowns, uncertainties, and challenges. Yet, such experiences have always been thought-provoking, fulfilling, and rewarding. I have been a seeker of new ways by challenging the status quo and doing my very best to orchestrate collective intelligence and efforts towards optimizing and enhancing user experiences, inclusion, security, and privacy with innovations that can effectively work and deliver on their promises. That's been my main driver, like something inside me constantly calling for change and positive impact.

I have also immigrated with my family twice so far, first to Dubai as a professional expatriate and then to Canada, pursuing my dream. Remember, you can only achieve to the maximum extent of the socio-cultural and economic boundaries and opportunities available to you to accomplish your vision. Look beyond the horizons with an open mind and be realistic about external limitations.

I've been privileged and humbly grateful to have had the opportunity to launch and promote various novel products to the mass market. These projects involved numerous moving parameters and details that needed to be discovered and addressed to succeed. These factors included establishing new standards, ensuring consumer centricity, assessing technology readiness, expanding market reach, evaluating implementation feasibility, and adapting organizational design, culture, and its readiness and willingness to embrace change, to name just a few.

It has been a very exciting and fulfilling professional journey so far, filled with challenges that turned into opportunities to learn and grow. Such initiatives have not always been well perceived. There may be a group of users who have been performing certain tasks in an old-fashioned, manual manner for a very long time. Not everyone in a cross-functional enterprise setup will understand or share the same vision. Effective stakeholder management and identifying the value and tangible benefits that such innovations can bring to each individual, team, or organization will significantly increase the likelihood of your success.

The lessons learned and the trajectory of prior technological innovations are essential inputs for understanding and effectively speculating on how and where the new wave of technology will land. I would like to share some examples and lessons learned from my adventures that we launched in Canada, the USA, and worldwide, which can be carried over to the context of AI transformation:

World's First-Generation Cloud-Based Mobile Payment in the 2000s

This innovation marked the beginning of a significant shift in the entire ecosystem from mobile operators' TSM-based dominance to the banking, fintech, and BigTech industries. It was quickly followed by proactive takeovers by original equipment manufacturers' (OEMs') wallets (e.g., Apple, Google, Samsung, etc.). This sub-segment of the market is forecasted to be worth $4 trillion by the 2030s.

Lessons Learned
New technology can disrupt business models, transforming them into new formats with significant financial and business implications. This often renders major incumbents redundant while making new entrants prominently relevant. Such dynamics are fundamental to the business landscape.

AI Gateway for Smart Home Devices

As the popularity of smart home connected devices and AI engines (e.g., Alexa, Apple, Google) surged, I designed a model that aggregates interfaces with all major AI engines and smart home or mobile devices, making it device agnostic. This model delivers secure banking and payment services loaded with next-generation cybersecurity, biometric authentication, and authorization, as well as technological and compliance features and use cases. Imagine being able to securely interact with AI in a B2C model.

On the other side of the equation, considering the B2B aspect, banks, fintechs, or financial institutions can offer and expose their services in smart homes and devices, powered by major AI engines in the world, in a device-agnostic manner, by integrating with a single interface only. Instead of having to integrate with all AI vendors and manufacturer interfaces, they can connect to this AI gateway, and we'll handle the rest for them in the most economical way, which is also faster to market. I was fascinated by this new technology and building some ideas around it. Then when my manager mentioned he had bought an Alexa and asked what we could offer in that space, I promptly designed the entire concept, from technical design to business model and market analysis, covering the whole nine yards afterward.

Lessons Learned
The first key point is to start with conceptualization, even at a very abstract, vague, and high level. Then, build it top-down, adding details as you progress.

Another key point is to break down the problem into smaller, manageable, and relatable pieces. For example, I separated the B2B2C model into B2B and B2C components to better understand the pain points and the value that the solution would bring to each user group.

Lastly, invention often involves combining two or more concepts that already exist. Always keep an eager eye on new technologies in other areas.

AI-Powered Biometric Authentication, Verification, Onboarding, and Digital Identity

Years before biometric and facial recognition features were introduced on handsets and mobiles, we were already leveraging this technology. The banking industry, however, was a bit late in adopting these innovations, allowing BigTech and OEM device manufacturers to capture the majority of the market share, similar to the mobile payment scenario. Digital identity represents another intersecting niche in this use case, forecasted for exponential growth. It is gradually rolling out across US states, with a current adoption rate of 1%. The Pan Canadian Trust Framework also has significant potential for growth and further solidification.

Lessons Learned
This is yet another example of disruptive change, transforming cross-industry ecosystems. Additionally, it is very important to take industry characteristics into account. In this case study, the banking industry relinquished the major potential of digital identity and its future prospects to other sectors, such as OEMs, fintech, and BigTech. Therefore, it's crucial to explore unconventionally different verticals and use cases when introducing new technology.

Automation and Digitization of Fulfillment, Supply Chain, and Inventory Management

This initiative transformed critical financial infrastructures and the payment ecosystem in Canada, covering 100% of Canadian personal and commercial payment cards and instruments. In this case study, let's discuss skeptical users. I recall socializing the rollout of this new service with heads of customer service and other key functions at a major tier-one bank in downtown Toronto. Initially, they adopted a defensive approach and were very skeptical about the new process. However, after I demonstrated the solution and highlighted how it could eliminate much of the manual work and hectic back-and-forth follow-ups by digitizing this part of the mass-scale industrial process, their perspective shifted. Not only would this free up their time for a warm coffee, creative work, or any preferred activity, but it would also securely safeguard the sensitive information previously sent over emails and files. At the end of the session, they all had big smiles on their faces and supported me in championing the rollout at their tier-one bank. The following week, another major financial institution, not one of my clients, experienced a breach where hundreds of thousands of Canadians' banking and personal data were compromised in exactly the same scenario I had cautioned against. An Excel file containing all personal data was breached. Had they been my clients, they would have avoided this exposure and also freed up more time and bandwidth for their employees.

Lessons Learned

User convenience and security are no longer mutually exclusive or a tradeoff; they can both be achieved through the right design of technology and processes. The best ethical and practical practice is always to train and empower employees to utilize new technologies versus replacement. Another lesson we can learn and excel in is the importance of emotional intelligence in driving strategic changes, such as digital transformation, and leading diverse groups of stakeholders toward a shared mission and vision. The key point is to understand your own motivations first. Then, fully comprehend the motivations of other groups, especially the skeptics. Finally, take the environmental context into account and formulate how both objectives can be achieved through shared goals and collective intelligence, effort, and success.

API-Driven Consumer-Driven Open Banking Cloud Services

A decade earlier, we made products ready for the emerging regulations and trends in open banking.

Lessons Learned
Regulation and compliance are core drivers of technological innovation. Additionally, it's essential to always monitor global and regional external factors, such as market trends, legislation, or best practices, and examine their common sense and potential impact. Pragmatically, prepare for innovation, plan exit scenarios, and experiment with what's upcoming much sooner rather than later.

Real-Time Payment and the World's Largest Real-Time Banking Services

These were introduced more than 20 years before their debut in Canada and the USA this year. Some technologies are driven by regional market characteristics and regulations.

Lessons Learned
This is very similar to the previous case study. Always monitor other global and regional markets. If a use case, regulation, or technological advancement makes perfect sense and can solve local problems, it's worth learning from and will likely materialize at some point. Therefore, it's crucial not to overlook the importance of looking back to learn from other regional markets or verticals. Localize your innovations by integrating new technological flavors and use cases.

SaaS APIs in Payment Issuance and Omni-Channel Digital Optimization

These were introduced at a time when legacy processes had been the established norm for decades. I presented my concept at our corporate Hackathon in Berlin to the group board and C-suite, securing their buy-in to address the API initiative at a strategic level. Subsequently, I was assigned to lead this initiative and successfully launch it in the following years. I focused on enhancing user convenience at every step of the user journey and user preference, enabling the "anytime, anywhere, any device or channel" concept. This portfolio of products played a crucial role in positioning the company I worked for as one of the top two technology providers in the modernized payment issuance category, as recognized by Juniper Research.

Lessons Learned
The very core element and mindset of innovation and practical transformation is rooted in questioning assumptions and challenging the status quo.

Another critical lesson I learned is the importance of having C-level or senior executive champions to support transformative strategic initiatives, as this significantly increases their likelihood of success. Innovation from within, can often elevate brand positioning, leading to substantial long-term business value and potential. Patience and persistence are key, by iteratively refining, optimizing and segmenting your vision and strategy into short-, mid-, and long-term goals, ensuring alignment with business, operational, and technological feasibilities and possibilities.

I think the aforementioned adventures serve as a testimony to my dedication to technological advancements for the greater good, making a positive impact on people's daily lives and interactions. Now that I look back, I'm deeply grateful for the opportunity to build lifelong connections and achieve collective successes with top experts, renowned brands, and influential stakeholders around the world. It has been an honor to contribute my part to the vast ocean of technological breakthroughs, collective knowledge, and expertise.

The Journey of Artificial Intelligence Evolution: Technology S-Curve

Contemplating the nature of AI: how much of it can be blended with human activities? How reliable and mature will it be for each use case, in due course and along the AI technology and market trajectory timeline? These were the initial questions I asked myself when I was tasked with leading a course on AI for an audience keen on the intersection of AI, fintech, paytech, and cybersecurity innovations, exploring new horizons of use cases and possibilities in other major industries. To answer these two questions, it is crucial to understand AI's technological maturity.
One of the well-known frameworks for measuring such a trajectory is the technological innovation S-curve, or simply S-curve, to analyze,

monitor, and measure a technological advancement and adoption rate within the school of the Bass model, which has progressed over decades. From this point on, we'll refer to it as the S-curve. This concept is based on Sigmoid's mathematical function. Numerous factors at various levels influence and shape the trajectory of the curve. A simplified version is illustrated in the figure below, categorizing user groups based on how early they adopt AI technology:

Technology Adoption S Curve

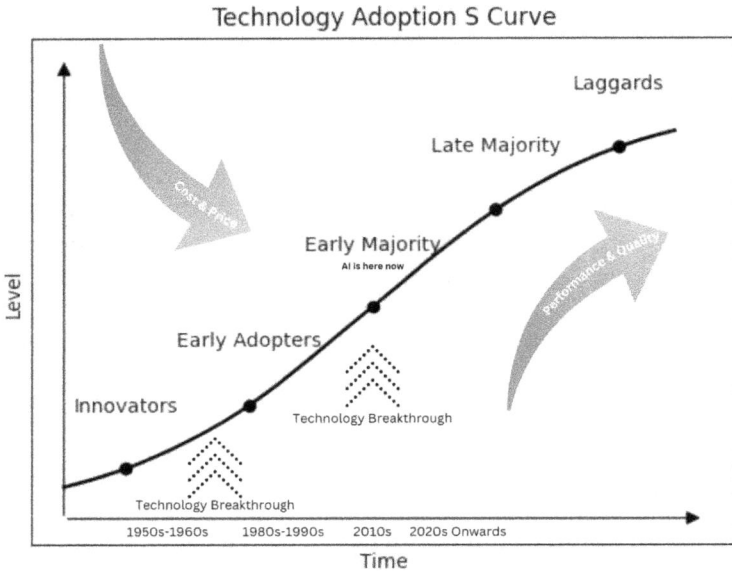

This model captures the lifecycle of technological breakthroughs and innovations, from inception to widespread adoption. It qualifies—and can potentially quantify and correlate—a wide range of technical and business parameters, including cost, revenue, emerging regulations, product performance, quality, market adoption, demographics, and more. It's a straightforward graph that allows technological innovation's performance and adoption rate to be analyzed based on its position along the S-curve timeline.

The key success factor, catalyst, and focal point is a technological breakthrough that makes the innovation accessible and propels it upward on the S-curve. In some cases, major breakthroughs can trigger a series of S-curves within the same technology, each emerging and elevating as the innovation evolves and matures. At the beginning of the S-curve, barriers, unknowns, and uncertainties are typically very high, which drives up the cost of developing or the price of adopting or obtaining a technological innovation. In other words, the technology is not yet mature and remains primarily at the stage of research, experimentation, proof of concept, or prototyping.

As AI technology has matured over time, these barriers have diminished, enabling broader accessibility for third parties and startups, more feasible consumer journeys, and growing use cases with potential integration across various industries and applications. This has resulted in exponential improvements in technological performance and quality.

Although AI has recently become a major hype, its journey has been a long one. Here's a high-level timeline: AI research and early models began with the initial development of neural networks and foundational algorithms in the 1950s and 1960s, marking the "innovators phase." Pioneering projects, such as ELIZA and the Perceptron, were significant milestones. Then, during the 1980s and 1990s, in the "early adopters phase," with chess playing computers and expert systems, AI began to see more practical applications, such as IBM's Deep Blue, the first computer to defeat a world chess champion, Garry Kasparov. The reintroduction and advancements in neural networks during the late 1990s and early 2000s marked significant progress in machine learning techniques, leading to the revival of neural networks.

With products like Apple's Siri, introduced in the 2010s, home assistants brought AI into households, marking its mainstream impact in consumer products and elevating it into the "early majority phase" on the S-curve. Autonomous vehicles and healthcare applications began showcasing trust in AI. Regulatory, legislation, and ethical frameworks began gradually solidifying and recalibrating globally. This is also the phase that I started my involvement in and new product introduction with AI.

AI is entering the "late majority phase" with widespread business adoption integrated into daily processes, from fraud detection in finance to demand forecasting and personalization in retail. AI-driven robotics are transforming manufacturing, logistics, and service industries, enabling advanced automation. During the "laggards phase," full automation in traditional industries is expected. Already-adopted AI systems will be further optimized, reaching new highs in performance, quality, availability, and affordability.

Embracing the Future Together: The Intersection of Technology and Humanity

Stay vigilant and maintain an open growth mindset when utilizing technological advancements such as AI. Always incorporate lessons learned from previous projects, products, or personal experiences. Think back to the first time you used mobile devices, smart gadgets, the internet, or streaming services—technologies that made life easier or more enjoyable. Let technology serve you, your customers, teams, products, and processes. A human-centric and ethical approach toward creating positive impact is essential to navigate the AI journey and integrate it, as a complementary tool, into our daily lives and experiences. After all, it is humanity that creates technology to serve us better toward a bright future and greater good.

About the Author

Parham Emami holds a master's in management of technology from the University of Waterloo and a degree in computer engineering. He is a certified expert in PMP, agile and hybrid project management, product management, software programming, artificial intelligence, paytech, SaaS, cybersecurity, public speaking, and strategic innovation from Gartner's top-ranked global institutes.

A trusted advisor to the C-suites of preeminent global technology providers, Parham has pioneered strategic innovations over two decades. He has launched over 200 projects, next-generation products, and platforms, ensuring substantial business growth and robust digital transformation in Canada, the USA, and worldwide.

Also a multi-award-winning artist known as QuAnthem, recognized as a top national talent by CityTV, he has performed at iconic festivals and collaborated with graduate studies at Toronto Metropolitan and Northeastern Universities on his projects, blending art and technology in big data, advertising, digital marketing, branding, and management studies, showcasing his commitment to innovative, impactful work.

LinkedIn: @ParhamEmami

CHAPTER 12

ETHICS OF ARTIFICIAL INTELLIGENCE: DEALING WITH CHALLENGES AND OPPORTUNITIES

By Craig Fulton MSc, Fin, CITP, MBCS
Technologist, Consultant, Speaker, Author
Singapore, Singapore

> *Whether we are based on carbon or silicon makes no difference; we should each be treated with appropriate respect.*
> —Arthur C. Clarke

Technological change opens new frontiers so rapidly with artificial intelligence that it alters the nature of industries, progresses human capabilities, and brings a new shape to the civilised world. As this technology permeates every part of our society, it will bring a host of ethical challenges that will have to be overcome.

The Future of Work: New Jobs, Responsibilities, and AI

Perhaps the most dire prediction about AI is its effect on the future of work. Every day, the rapid rate of adoption of AI-based systems is automating thousands of tasks as the human workers are laid off, inspiring the phenomenon of technological unemployment.

According to Pachegowda (2024), this transition requires a fundamental reimagining of the roles and responsibilities of humans and machines and the working population's upskilling, reskilling, and cross-skilling.

The Ethics Conundrum: Balancing Innovation and Accountability

The more AI percolates into every aspect of our lives, the more its ethical dimensions become progressively more complex. Achieving a balance between driving innovation and ensuring accountability is a considerable challenge due to issues with transparency and explainability of the AI system itself, the possibility of bias and discriminatory practices, and ethics involved when applying AI in sensitive areas like healthcare, criminal justice, and financial services. Addressing these challenges requires a collaborative process in which policymakers, business leaders, and ethicists work out robust ethical frameworks and governance structures to protect human well-being and the safety of society as a whole.

Bias: Mitigating Discrimination in Algorithmic Decision-Making

One of AI's most important ethical challenges is algorithmic bias, in which AI systems reflect and sometimes amplify existing cultural biases (Ferrara, 2023b; Ntoutsi et al., 2020). It may come in many forms, such as gender, race, or socio-economic status, and lead to discriminatory outcomes in areas including but not limited to hiring, credit scoring, and criminal justice.

Given this risk, researchers and practitioners have developed various fairness-aware machine-learning techniques, diversity in data, and human-in-the-loop oversight. Such works are represented by, among others, the researchers Madaio (2020), Ferrara (2023a), Ntoutsi (2020), and Khan (2023). Nevertheless, the challenge remains a major concern that requires relentless vigilance and continuous collaboration along the path of equipping all AI systems with fairness and non-discrimination capabilities.

Transparency and Explainability

Another important ethical consideration involves the transparency and explainability of the AI system. Most AI models are "black boxes," meaning nobody can reach conclusions. Without this transparency, trust will rapidly wear away, accountability is often hindered, and there will be significant grounds to question the fairness and integrity of AI-driven decision-making.

The reason so much effort has been invested in researching means of enhancements that may improve the interpretability and explainability of an AI system, including the application of explainable AI, is because of this very problem. The aim is for AI's decision-making functioning and processes to be made transparent and understandable. This would provide greater transparency to give credence to public trust, to allow proper oversight, and to ensure AI is deployed and managed ethically and responsibly.

Privacy: Safeguarding of Personal Data in the Digital World

One glaring implication of these technologies that have rapidly and un-precedentedly permeated daily life is the grave dispute over personal data privacy and security. As these AI systems have become increasingly so-phisticated, capable and efficient at collecting, processing, and analysing huge quantities of data, so too have threats from external intrusion, data breach, and misuse of sensitive information, with the risk and severity of these threats rapidly escalating.

In response, policymakers and industry actors have presented various frameworks and regulatory mechanisms for data governance that protect the privacy of people and promote ethical values, such as the General Data Protection Regulation (GDPR) in the European Union. But with the arrival of these AI-enabled innovations, assurance that our personal data will remain private is one of the things we need to be conscious about and to work together on.

Driverless Vehicles: Navigating the Ethical Road

Here come specific ethical challenges associated with integrating AI into autonomous cars, from issues of responsibility in the event of a car accident to tracing ethical justification for a choice between lives in cases of inevitable collisions to even how all this affects transportation equity and access. These ethical issues could be answered through the active cooperation and collaboration of policymakers, car manufacturers, and AI professionals to draft all-encompassing frameworks that ensure the safety, equity, and protection of human life.

Healthcare: A Better Diagnosis and Treatment

AI integrated into healthcare systems will potentially revolutionise the whole concept of diagnosis and treatment. AI advances the enormous tasks of analysing large amounts of patient data, identifying patterns, and making personalised treatment recommendations with better accuracy using AI-powered tools.

In contrast, the use of AI in healthcare offers a more prominent set of ethical discussions involving data protection concerns, issues with bias, and risks introduced from replacing or mitigating roles played by healthcare professionals. The benefits of AI in health can be realised, coupled with mitigating these ethical risks, by having robust frameworks and guidelines that anchor patient privacy, fairness, and preservation of the human-centricity of healthcare.

AI applied to health holds great promise, but many ethical concerns accompany it. Hence, good governance frameworks, continuous surveillance, and commitment to principles for ethics become of prime importance to ensure that the use of AI in health will translate into better outcomes for patients without violating the rights of a person and human-centred values.

The Environment: Sustainable Solutions or Unintended Consequences?

Rapid advances in artificial intelligence have created colossal opportunities and significant challenges concerning environmental sustainability. On one hand, AI-powered technologies may completely change how we manage and optimise natural resource use, energy, and waste reduction. Some firms have used AI to enhance their environmental performance by implementing energy management strategies, using natural resources, reducing waste, and monitoring carbon. They have created sustainable enterprise business practices. At the same time, however, it should be noted that most AI processes have environmental consequences in that most are done using energy-intensive training and running of models that can have high carbon footprint emissions. The possible long-term environmental damage caused by AI needs to be managed, and strategies to ensure sustainable deployment should be instituted now.

The Educational System: Transforming Learning and Teaching

AI has changed the face of higher education learning and teaching, and along with exciting opportunities, many serious ethical concerns have also arisen. The prospect of automatic scoring, fast content analysis, and personalised feedback has become feasible with AI technologies. They have enhanced the assessment experience to provide deep insights into students' performances.

On the other hand, the deployment of AI in education has raised critical concerns about validity, reliability, transparency, fairness, and equity. Issues such as algorithmic bias and opacity within AI decision-making processes may create a risk of perpetuating inequality and may impact assessment outcomes. Many of those challenges are being addressed through guidelines developed by various stakeholders to ensure that AI is used in education ethically and responsibly.

Criminal Justice: Ensuring Fairness and Due Process

The application of AI to the criminal justice system has raised several questions about fundamental fairness and due process. AI-driven tools, like risk assessment algorithms, have helped determine decisions over bail, sentencing, and parole. Theoretically, this makes an otherwise subjective process even more objective, but, as it happens, it opens up a vista for biases to creep in and perpetuate existing inequities within a system.

All these ethical considerations raise serious demands in the perspective of the development and deployment of AI systems regarding bias testing, strictness regarding transparency of the decision-making process, and ongoing monitoring for fairness and accountability.

Financial Services: How to Balance Risk and Opportunities

AI technologies are among the early adopters of financial services as an essential tool to enhance decision-making and fraud detection and give customer experiences a personal touch. Even though significant competitive advantages could be reaped regarding efficiency, speed, and automation, using AI in finance raises some serious challenges.

These would include transparency, interpretability, equity, accountability, and trust issues. The "black box" operation of some AI models, without interpretability, may be used as a basis for arguments that these would result in biased or poorly based decisions. Debates over data protection and information security are raised with AI in financial services.

AI and Social Media: Content Regulation and Misinformation Diffusion

The proliferation of AI-powered social media has increased the rate of misinformation and disinformation, among other forms of hurt. More recently, AI algorithms have also been increasingly used for content curation and recommendation, boasting some of the most sensational, polarising, or factually incorrect pieces of information.

These ethical challenges will only be overcome through multi-pronged processes that include the development of good content moderation policies, increasing transparency in algorithmic decision-making and encouraging greater collaboration by social media platforms in cooperation with policy thinkers and civil society regarding regulating AI use in this area.

National Security: Walking the Line Between Technological Progress and Ethical Issues

AI integrated into the national security and defence structures changed everything, from how surveillance is conducted down to making decisions and unmanned war weaponry. Such an advancement might improve national security, but such developments have raised several serious ethical challenges. One, the development and deployment of killer robot systems have sparked worldwide debates on the morality of giving machines the power to decide life and death. Policymakers and military leaders must consider the competing imperatives of technological advancement versus human agency, dignity, and accountability.

The Workforce: Build Skills for Future Readiness

AI being used in a number of spheres has raised many eyebrows as far as implications to the workforce. Automation through AI may remove workers from some job functions, which might result in unemployment and, hence, would require training and reskilling of the workforce.

This issue requires a partnership between the employer, educational institution, and policymaker to ensure that the workforce is suitably prepared to respond to the challenges of working and living in an AI-dominated economy. In preparing workers for the shifting job market, proactive investment in lifelong learning and developing skills to meet the challenge is necessary (Faishal et al., 2023).

Artistic Expression: Preserving Human Creativity

The rapid and numerous improvements in AI-generated art raise quite an active debate within the artistic community. Yet, while AI-driven tools open up entirely new dimensions and possibilities of creative expression, all this needs to be weighed against the displacement of human artists and the devaluation an artistic work may entail as being created by human hands. This calls for a balancing act that respects the values of both the artistic expression of people and the creations of AI without eroding the hard-won rights of artists and any loss of human ingenuity.

Ethical Issues in AI and Robotics: How to Interact Between Humans and Machines

As more and more areas as far-reaching as health and manufacturing have started applying the power of AI and robotics, several ethical issues have cropped up. These include security, reliability, and transparency of AI-powered robots, human job displacement, and developing human-machine interfaces. Ethical frameworks and guidelines should be developed in collaboration with policymakers, researchers, and industry players to protect human well-being and agency and ensure the design and deployment of AI-powered robotics is done responsibly.

The Global Economy: Charting Consequences in Geopolitics

AI is fast becoming a transformative technology, which holds immense consequences for the world's economy and geopolitics. The current disparities and power plays between and among countries and regions are accentuated by the uneven availability of AI, leading to more competition and conflict. The geopolitical challenges thrown up by the rise of AI, thus, demand an inclusive, multilateral approach toward responsibly and equitably.

Ethics and Morality in the Light of Artificial Intelligence: The Setting and the Practice of Moral Principles

Setting and implementing robust ethical frameworks has become urgent as AI capabilities increase. Some basic questions—such as how AI systems should align with human values, where moral responsibility might be attributed, and how different ethical principles will be balanced—need to be answered. Besides those from policy thinkers, ethicists, and AI researchers, ongoing efforts are essential for developing comprehensive ethical guidelines and governance mechanisms concerning the responsible development and deployment of AI technologies.

Artificial Intelligence and the Human Experience: Enhancement or Replacement?

The increased application of AI in more aspects of human life raises fundamental questions about the human experience. Where, on the one hand, AI expands the ability of humans and helps improve the quality of life, it conversely faces such concerns as the accusation of replacing human agency, dissolving the boundaries between human and machine, and carrying in itself the prospect that AI may alter or take away the essential elements of being human. Issues like these require nuance and thoughtfulness not to harm the protections around human dignity, personal agency, and uniqueness associated with the human condition.

Governance: Creating Effective Policy and Regulation

The rapid speed of recent developments in AI is highly challenging for policy actors and governance bodies to keep up with. They build comprehensive regulatory frameworks at rates that lag way behind. As such, the demand is increasing for the creation of appropriate policies and regulations concerning the responsible development and deployment of AI systems.

These are data privacy, algorithmic transparency, and mechanisms of accountability and oversight, whose enforcement measures were set that could mitigate the risks associated with this AI technology. Numerous AI researchers and thinkers have propagated this (Bostrom & Yudkowsky, 2014; Hernández, 2024; Fukuda-Parr & Gibbons, 2021; and Krafft et al., 2019).

Conclusion

As we have learnt in this chapter, the implications of artificial intelligence are multifaceted and dynamic. The rapid development of technology in AI has brought together a wide range of opportunities in our lives, from a revolution in healthcare to new ways of working and learning. However, these technological breakthroughs have also introduced many ethical issues that must be addressed with utmost care and diligence.

AI bias is one of the significant challenges that may lead to increased discrimination in society. It will help build public confidence in AI if the designing and implementation of AI aims for reductions in bias and increases in fairness necessary to uphold fundamental human rights. The third concern relates to the problem of transparency and explainability of AI decision-making. AI is both an object of growing complexity and also an object of growing incomprehensibility and, hence, an unaccountable system (Bostrom & Yudkowsky, 2014).

Another critical area of concern is the effect of AI on privacy and personal data. As AI-powered technologies penetrate all dimensions of human life, the demand for privacy protection as a fundamental human right to be protected from misuse or extortion of personal data becomes louder. In addition, AI ethics is concerned with more than individual rights; it has broader societal implications as well. Another area where challenges associated with the ethics of AI arise is in the international development space.

The challenges around the use of AI are extensive but serve as an opportunity for positive change. Through a trans-disciplinary engagement with experts from AI, ethics, and policy, AI systems that work toward enhancing human wellness, advancing social justice, and protecting the

Earth can be created. However, time is of the essence. The ethical issues around AI need to be addressed now.

About the Author

Craig Fulton is an accomplished technologist, dedicated consultant, inspiring executive coach, captivating speaker, and insightful author.

Email: craig.fulton@ibm.com

CHAPTER 13

BLUEPRINTS AND BEYOND

By Jeremy Kofsky
Founding Member, The AI Circle; GenAI Strategist
Washington DC, USA

What do you want?
—The Notebook

Whether in military or business strategy, all great plans start with a clear goal. In military terms, this is known as the commander's intent. It clarifies the essential objective so everyone knows the critical action to ensure mission success even if everything else fails. This clarity allows teams to adapt to challenges and stay aligned with their "North Star." Without this, in a modern corporate setting, all the artificial intelligence in the world will not support an organization's transformation or optimize its efficiency. With this, AI-enhanced operations can keep strategies adaptable and aligned with the vision. From the battlefield to the boardroom, having a clear picture of success from the start guides every decision, simplifying complex tasks and steering all efforts toward a common, defined end. So, no matter what the arena, remember: real success starts with a clear destination.

As strategic plans unfold, clear vision sets the stage for integrating sophisticated tools like AI. AI is not just a tool but a transformative force to redefine the landscape of strategic operations. By automating complex data analyses, optimizing logistical operations, and predicting future trends and scenarios, AI enables organizations to leapfrog traditional limitations of speed and accuracy, thereby redefining the strategic landscape.

The thoughtful deployment of AI takes advantage of its ability to dovetail seamlessly from and with the strategic vision. It amplifies the capabilities of human teams, allowing them to transcend conventional performance metrics by offloading routine and computationally intensive tasks to machines. This symbiosis between human strategic oversight and AI-driven execution ensures every layer of the organization is optimized for efficiency and aligned with the strategic goals, demonstrating the practical benefits of AI integration.

Thus, the pivotal question in strategic operations—"What is being done, and why?"—is enriched by AI's capacity to provide deeper insights and foresight, enabling leaders to conceive and execute strategies with unprecedented precision and adaptability. This dynamic interplay between vision and technology ushers in a new era of strategic operations where decision-making is data-driven and intuitively aligned with long-term objectives.

At its essence, strategic planning demands a clear, actionable vision propelling all organizational activities toward a well-defined goal. Simon Sinek emphasizes the importance of starting with "why," stating, "People don't buy what you do; they buy why you do it. And what you do simply proves what you believe." This principle is vital for strategic planning, where clarity of purpose and intent critically influence the effectiveness of outcomes.

Effective strategy merges this vision with practical execution. It commences with a comprehensive assessment of the current landscape, blending an understanding of the environment with foresight into potential future changes and challenges. In a corporate context, this might involve analyzing market trends, consumer behavior, and technological advancements to pinpoint opportunities for innovation or areas of potential disruption.

A notable example of military precision applied to corporate strategy is evident in how companies approach market entry or product launches.

Much like how military strategists conduct detailed reconnaissance to customize their operations, companies analyze market data to discern consumer needs and competitive gaps. The strategy is meticulously crafted to exploit these insights, ensuring each tactical maneuver is aligned with broader business objectives.

Colin Powell once remarked, "There are no secrets to success. It is the result of preparation, hard work, and learning from failure." This sentiment underscores the meticulous nature of strategic planning where scenario planning and risk assessment are pivotal. This methodical approach, akin to military war-gaming, involves simulating various scenarios to fully understand potential risks and to develop effective counter-strategies.

By rigorously exploring these possibilities, leaders can anticipate challenges and adapt their strategies proactively, ensuring the organization remains resilient and agile. This foresight is crucial not only for maintaining operational stability but also for capitalizing on dynamic opportunities, thereby enhancing the strategic agility of the organization.

Former CEO and business strategist Jack Welch reinforces this perspective, stating, "Change before you have to." In strategic planning, this proactive adaptation is facilitated by continuous environmental scanning and scenario analysis, enabling organizations to stay ahead of the curve and preemptively adjust to market forces and competitive pressures.

Ultimately, strategic planning is about aligning vision with execution—crafting a roadmap to not only anticipate the expected but also being resilient enough to withstand the unexpected. As military and business environments continue to evolve, the principles of strategic planning remain constant, providing a stable framework for navigating the complexities of modern operations.

AI-Enhanced Operations

AI is revolutionizing how organizations execute their strategic plans, providing tools to perform tasks and analyze and predict outcomes from complex datasets. This shift from human-led to AI-assisted operations marks a significant enhancement in operational capabilities, dramatically improving speed, accuracy, and adaptability. As a strategic plan is crafted and grounded in a clear vision and detailed analysis, the integration of AI is pivotal in elevating operational efficiency and effectiveness. AI is not a replacement for the strategic foundation established through thorough planning; it is a complement and enhancer, automating and optimizing processes to achieve predetermined goals with unprecedented precision and speed.

The initial impact of AI on strategic operations is often seen in its ability to manage large volumes of data with ease. Traditional data analysis methods, while reliable, require substantial human effort and are prone to errors, especially under time constraints. However, AI systems process these datasets quickly and accurately, offering previously unattainable strategic insights within such tight timeframes. This capability enables organizations to make more informed decisions quickly, providing a critical advantage in both corporate and military settings.

For example, predictive analytics can forecast market changes or identify potential security threats before they materialize, enabling preemptive actions that align with the strategic goals outlined during the planning phase. This enhances an organization's responsiveness and solidifies its proactive stance in managing potential challenges.

Tiarne Hawkins, a noted AI strategist, underscores the transformative impact of AI, stating, "AI's predictive capabilities are transforming how we approach strategic planning. By analyzing past and current data, AI helps us accurately forecast future trends and scenarios, allowing for proactive rather than reactive strategies." Hawkins highlights the importance of AI in strategic contexts, emphasizing its role in enhancing decision-making processes and ensuring organizations can adapt swiftly to environmental changes.

AI models learn from data iteratively and improve over time, enabling them to uncover patterns and insights that human analysts might miss. This capability allows for strategic adjustments both timely and data-driven, significantly enhancing operational effectiveness. Moreover, AI automation of routine tasks frees human resources to focus on more complex strategic considerations and innovation. This redistribution of tasks boosts productivity and fosters creativity, as employees can dedicate more time and energy to tackling complex problems and developing innovative solutions.

In conclusion, the integration of AI into strategic planning is not merely an enhancement but a fundamental evolution in the operational frameworks of modern organizations. It empowers businesses and military operations alike to navigate today's complex landscapes more effectively, transforming challenges into opportunities and data into strategic foresight. This pivotal shift in operational paradigms highlights AI's role as an indispensable strategic partner in designing a future to meet or exceed stated goals and aspirations.

—

As we delve deeper into the specific applications of AI in strategic operations, various techniques come to the forefront, including the following three:

Robotic Process Automation (RPA)
Robotic process automation (RPA) significantly streamlines operational efficiency by automating routine and repetitive tasks. Within strategic contexts, RPA is invaluable for executing logistical operations, processing financial transactions, and managing extensive data entry processes. By automating these foundational tasks, RPA not only reduces the potential for human error but also liberates skilled personnel to concentrate on more complex and critical strategic decision-making and oversight tasks. This shift in focus from mundane tasks to higher-level strategy enhances the effectiveness of planning and execution across business sectors, allowing organizations to be more responsive to market dynamics and operational demands.

Retrieval-Augmented Generation (RAG)
Retrieval-augmented generation (RAG) is a sophisticated AI technique that dramatically enhances decision-making processes. It achieves this by retrieving a vast array of information from extensive databases and augmenting this data with generated content that is contextually relevant. In strategic environments, RAG is especially beneficial for synthesizing historical data and contemporary case studies to support informed decision-making. By providing a rich, contextually enhanced understanding of situations based on accumulated knowledge, RAG allows strategists to craft plans that are not only responsive to current conditions but also deeply informed by historical precedents. This capability is crucial for developing strategies that are robust, adaptable, and finely tuned to the nuances of the operational context.

Intelligent Agents

Intelligence agents represent a leap in AI technology, offering robust capabilities for autonomous operation within predefined parameters. These AI-driven agents are programmed to manage and execute a wide array of tasks independently, from monitoring supply chain dynamics to making real-time tactical adjustments in both military and business settings. The autonomy of intelligent agents allows them to react swiftly to changes, maintaining operational agility and alignment with strategic objectives even under rapidly evolving conditions. Their deployment can significantly enhance the operational responsiveness of an organization, ensuring that strategic goals are pursued efficiently and effectively, even in complex and fluid environments.

Transitioning from Theory to Practice: A Strategic Planning Framework

As we look towards the significant impact of artificial intelligence on strategic operations, it's essential to establish a clear method for its integration. A strategic planning framework offers a detailed guide for effectively harnessing AI within your organization. This approach ensures AI fits seamlessly into existing structures and aligns perfectly with strategic goals, enhancing overall effectiveness and efficiency.

Start with the End

Begin by defining the ultimate objectives and overarching narrative of your strategic initiative. This ensures every AI-enhanced activity is directly aligned with the organization's primary goals. Consider using AI to analyze historical data to better define these goals based on past successes and challenges.

Orient on the People

Identify key stakeholders and their roles within the strategic plan. Evaluate how AI can be tailored to augment their capabilities and improve workflow efficiencies. AI can also provide personalized training and development programs based on individual performance metrics and learning paces.

Achieve the Outcomes

Leverage AI to establish clear, measurable outcomes and objectives that support the broader strategic narrative. AI tools can continuously track these metrics, providing real-time feedback, allowing agile adjustments to strategies and tactics.

Support the Process

Leaders should foster an environment encouraging innovation and adaptability. They must also proactively manage change by communicating benefits and involving stakeholders while continuously evaluating AI's impact to ensure it aligns with the organization's strategic objectives.

The Road Forward: Strategic Implications

Integrating artificial intelligence into strategic planning is not merely a technological upgrade but a transformative shift redefining how organizations approach their operations. AI enhances the capability to act on well-founded strategic plans by providing tools like robotic process automation (RPA), retrieval-augmented generation (RAG), and intelligent agents to quicken the decision and execution cycle. These technologies extend human capacity, allowing for more precise execution and adaptive responses to complex challenges.

The fusion of AI with strategic planning creates a synergy where each element strengthens the other. Strategic foresight informs the deployment of AI tools, ensuring that their application is aligned with clear, purpose-driven goals. Conversely, AI brings a level of efficiency and insight previously unattainable, enabling organizations to leverage their strategic plans in ways that continuously innovate and adapt.

Looking ahead, the role of AI in strategic operations is set to expand, with advancing technologies offering even greater opportunities for innovation. As AI becomes more sophisticated, its integration into strategic planning will likely become standard rather than the exception. Organizations that can effectively harness this potential will not only stay ahead in their respective fields but also set new benchmarks for efficiency and strategic success.

In closing, we return to the essential question framing our approach to love, life, and strategy: "What do you want?" This query serves as both a philosophical pondering and a directive to envision and create a future optimizing the utility of tools at our disposal. As leaders and strategists, we must adapt to the evolving landscape of AI and actively shape it in a way that furthers our collective goals. By doing so, we ensure AI serves as a catalyst for strategic excellence, transforming how we operate and, ultimately, how we succeed in an increasingly complex world.

About the Author

Jeremy Kofsky is an accomplished senior consultant in AI with a distinguished military background, including 20 years of service in the United States Marine Corps, where he was a distinguished honor graduate of Expeditionary Warfare School and a Brute Krulak Scholar as an enlisted Marine. During his military service, Jeremy played a critical role in operations requiring precise decision-making and the integration of advanced technologies. His expertise in both military operations and AI has allowed him to bridge the gap between defense initiatives and cutting-edge technology, enhancing mission success and operational efficiency.

With a deep understanding of AI and its practical applications, Jeremy has guided organizations in implementing AI-driven solutions addressing complex challenges across industries. His work extends to business development and strategic advising, helping companies navigate the evolving AI landscape. In addition to his consulting and military careers, Jeremy is committed to community initiatives that promote technological education and innovation, particularly within defense and security sectors.

LinkedIn: https://www.linkedin.com/in/jeremy-kofsky/

CHAPTER 14

AI—ANOTHER TOOL IN THE TOOLBOX

By Brandon Lester, CISSP, PMP
CTO, Director of AI
Honolulu, Hawaii

> *A good tool improves the way you work. A great tool improves the way you think.*
> —Jeff Duntemann

The rise of artificial intelligence has sparked a mix of excitement, curiosity, and concern. As with any transformative technology, the challenge lies not just in understanding its potential but also in recognizing its limitations and learning how to harness it effectively. For many, AI remains a buzzword—promising to change the world yet often misunderstood or misapplied. However, AI is not a magic wand that can solve every problem effortlessly. Instead, it is a powerful new tool, one that requires skill, judgment, and ethical consideration to use effectively.

To illustrate, imagine starting a home improvement project. A screwdriver is reliable but a slow and labor-intensive tool. Introducing a power drill transforms the same task into something faster, easier, and scalable. Yet, the power drill doesn't replace the need for skill; it requires training and understanding to be used safely and effectively. Similarly, AI offers a leap forward in capability, enabling us to work more efficiently and tackle challenges at scale. However, like the power drill, it is not a substitute for human expertise, creativity, or ethical decision-making. To truly harness AI's potential, we must learn to wield it wisely, understanding when and where it adds value. By doing so, we can unlock its transformative power to complement our own.

Understanding AI's Role

AI's strength lies in its ability to process vast amounts of data, recognize patterns, and automate repetitive tasks with remarkable speed and efficiency. It can analyze massive datasets, generate insights, and perform tasks that would be time-consuming or impossible for humans to handle manually. For instance, AI powers predictive analytics, helping businesses anticipate customer needs. Amazon, for example, collects browsing and shopping data to make personalized recommendations—sometimes even predicting what you might want to buy before you realize it yourself. In healthcare, AI-powered diagnostic tools assist radiologists by analyzing medical images to detect anomalies with high accuracy. These tools process information faster than humans, but they are most effective when paired with a doctor's expertise to provide context and ensure accurate interpretation.

Despite its strengths, AI has significant limitations. Its effectiveness depends entirely on the quality of the data it is trained on, a concept often summarized as "garbage in, garbage out." This highlights the critical need for high-quality, unbiased data. When the data used to train an AI system contains biases—such as favoring specific demographics or educational backgrounds—the system not only inherits these biases but can also amplify them. Take AI in hiring, for example: these tools are designed to streamline recruitment by identifying top candidates based on patterns in historical data. However, if that data reflects past discriminations, such as a preference for male candidates, the AI will replicate and even reinforce those biases, systematically undervaluing resumes from women or individuals from underrepresented groups, even if they are equally qualified. This leads to unintended and unfair hiring practices that undermine overarching goals.

To prevent these pitfalls, AI systems must be meticulously trained, continuously monitored, and iteratively improved. Transparency in how these systems operate is key to building trust, as is ensuring that they consistently produce fair and accurate results. Regular audits of both the data and the algorithms are essential to identify and address biases or inaccuracies. Without careful oversight, AI risks losing credibility with its users. Once trust is eroded—whether due to biased decisions, inaccuracies, or opaque processes—it becomes challenging to rebuild. Users begin to question the system's validity, which can ultimately render it ineffective. Properly managed, however, AI has the potential to enhance decision-making and drive progress, provided it is wielded with care and accountability.

Applications in Work

In the workplace, AI is poised to revolutionize businesses by enhancing productivity and streamlining operations. Software companies are rapidly integrating AI features into everyday tools, making them more powerful and user-friendly. For example, Microsoft Teams incorporates Copilot, which transcribes meetings in real time, generating a comprehensive summary of discussions, action items, and other key takeaways. Generative AI systems like ChatGPT serve as invaluable jump-start helpers, enabling workers to quickly acquire new knowledge by summarizing complex topics, offering contextual recommendations, and producing initial drafts of documents, presentations, or reports. This functionality reduces the cognitive load of tedious tasks, empowering employees to focus on higher-value work.

Rather than replacing existing tools and expertise, AI complements them, enhancing their capabilities and integration within larger systems. One such innovation is agentic AI, which is designed to act independently within defined parameters to achieve specific goals. Unlike traditional AI systems that require user input for each task, agentic AI can proactively analyze situations, prioritize tasks, and execute complex workflows with minimal human intervention. For example, in project management, agentic AI can monitor deadlines, reallocate resources as priorities shift, and communicate updates to team members—all autonomously. This proactively allows businesses to focus on efficiency while freeing employees to concentrate on strategic planning, innovation, and relationship-building. However, as agentic AI takes on more decision-making responsibilities, it is critical to establish clear oversight, ethical guardrails, and accountability measures to ensure its actions align with organizational values and objectives.

AI also excels at automating repetitive, manual tasks, reducing the time and energy spent on routine activities. AI-powered tools can organize calendars by analyzing user preferences, identifying open slots, and suggesting optimal meeting times based on participants' availability. Appointment scheduling, often a cumbersome back-and-forth process, can be delegated to AI assistants that handle the logistics seamlessly. Additionally, these systems can set personalized reminders, ensuring users stay on top of deadlines and commitments without needing constant manual input. This automation not only saves time but also reduces administrative burnout and cognitive fatigue, enabling individuals to focus on more strategic, creative, or meaningful tasks.

Moreover, the potential applications of AI in the workplace extend to improving decision-making and collaboration. AI-driven analytics tools can process vast datasets to uncover patterns and trends that inform business strategies. In collaborative environments, AI can analyze team workflows, identifying bottlenecks and suggesting process improvements. By optimizing both individual productivity and team dynamics, AI empowers organizations to adapt and thrive in an increasingly competitive and fast-paced landscape. Ultimately, AI's role in the workplace is not merely to replace human effort but to amplify it, unlocking new opportunities for innovation and growth while fostering a more efficient and harmonious work environment.

Applications in Life

AI has become a part of our personal lives, integrating into daily routines through everyday technology. From shopping chatbots that recommend products to virtual assistants like Siri and Alexa that provide instant answers, AI is constantly simplifying tasks and making life more efficient. Smart home devices go a step further, adapting lighting, temperature, and home security settings to user habits in their living environment. While these applications are helpful, they only scratch the surface of AI's potential to enhance personal growth and fulfillment.

AI integrations will truly shine as a personalized coach, offering insights, tracking progress, and delivering consistent motivation to help individuals set and achieve their goals. By analyzing user habits, preferences, and past behaviors, AI tools can craft realistic, personal, and actionable plans. Productivity apps can break down ambitious objectives into manageable tasks, prioritize them by importance, and create daily schedules aligned with personal routines. Fitness trackers set incremental health milestones, such as increasing daily steps or improving cardiovascular endurance, while providing real-time feedback to keep users engaged. Financial tools monitor spending patterns, suggest budget adjustments, and forecast long-term savings outcomes, helping users achieve financial stability. Acting as a virtual coach, AI adapts its guidance as progress is made, ensuring that goals remain relevant and achievable. This dynamic and data-driven support turns aspirations into tangible successes.

Beyond goal-setting, AI enhances communication and relationships by providing thoughtful, real-time guidance. Relationship coaching apps analyze tone, timing, and phrasing in conversations, offering suggestions to foster empathy and constructively resolve conflicts. For example, AI could suggest rephrasing an important email to ensure the tone and content are tactful, increasing the chances of receiving a constructive and positive response rather than provoking a confrontational reaction. This type of AI-powered coaching helps individuals strengthen personal and professional connections by promoting clarity, empathy, and understanding.

In education, AI acts as both a teacher and mentor, delivering personalized learning experiences that adapt to a student's needs. Platforms like Khanmigo (Khan Academy's AI tutor) adjust lessons in real time to match a learner's progress and skill level, making the process more engaging and efficient. AI tutors provide targeted assistance, helping students overcome challenges such as complex math problems or essay writing. Teachers, in turn, can use AI-generated insights to better support their students, identifying areas where additional guidance is needed. By tailoring educational support to individual strengths and weaknesses, AI enables students to learn at their own pace and build confidence.

Whether coaching us toward personal growth, fostering stronger relationships, or empowering educational achievement, AI is transforming how we navigate and enhance our lives. It doesn't just automate tasks—it helps us unlock our potential, serving as a trusted partner in the pursuit of a more organized, connected, and enriched existence.

The Importance of Choosing the Right Tool

Not every problem requires AI, just as not every home improvement project needs a power drill. While AI is a powerful and transformative tool, it is not a universal solution, and using it indiscriminately can create more problems than it solves. One of the most common pitfalls is over-using AI or applying it where simpler, more effective solutions already exist. For example, automating sensitive tasks, like handling customer complaints, can result in impersonal and inappropriate interactions. In such scenarios, the lack of a human touch can damage trust, relationships, and credibility.

The key to using AI wisely lies in understanding its strengths and limitations, applying it to tasks it is well-suited for, and ensuring human oversight to guide its use. Like any tool, AI is only as effective as the person wielding it, and its value lies in how thoughtfully and strategically it is applied. By recognizing when AI is appropriate and when human judgment and empathy are indispensable, we can unlock its potential without falling into the trap of overuse or misuse.

AI and Human Collaboration

AI is at its most powerful when it acts as a partner, enhancing human capabilities rather than replacing them. This synergy between human expertise and AI efficiency creates opportunities for greater innovation and productivity. In software development, for example, AI-powered tools like Copilot assist programmers by generating code snippets, offering suggestions, and automating routine tasks. This not only speeds up development but also reduces the cognitive load on developers, allowing them to focus on more complex problem-solving and creative aspects of their work. However, these tools do not eliminate the need for skilled developers. Human expertise remains essential to guide the AI, refine its outputs, and ensure that the final product meets both technical and ethical standards.

This symbiotic relationship underscores a critical truth: AI and humans, working together, achieve more than either could alone. In journalism, for instance, AI can process vast amounts of research and information to identify patterns, trends, and key insights much faster than journalists. Yet, it is the journalist's judgment and storytelling skills that transform raw details into narratives that resonate with audiences. AI provides the foundation, but it is human ingenuity that gives the work its meaning and impact.

To fully realize AI's potential, individuals must develop the skills to wield it effectively. AI literacy—the ability to understand, interact with, and evaluate AI systems—is rapidly becoming an essential skill across industries. Just as computer literacy became a fundamental requirement during the digital revolution, AI literacy is emerging as the key to thriving in the AI-driven future. Training and upskilling initiatives must prioritize equipping people with the knowledge and tools to harness AI responsibly. This includes understanding AI's limitations, ethical considerations, and best practices to ensure that its benefits are widely accessible and equitably distributed.

The implications for the workforce are profound. Jobs are not disappearing as much as they are evolving. While some routine tasks may be automated, new roles are emerging in fields like AI system training, maintenance, and oversight. For example, a teacher might use AI tools to track student progress, identify learning gaps, and recommend personalized learning paths. However, the teacher's role in mentoring, inspiring, and fostering critical thinking remains indispensable.

The key takeaway is that the integration of AI into the workplace and life is not about replacement but collaboration. By leveraging AI to enhance human strengths, we unlock opportunities for greater efficiency, creativity, and problem-solving, ensuring that technology serves as a tool for empowerment rather than a substitute for human potential. This balanced approach will shape a future where AI complements, rather than competes with, human capability.

Conclusion: A Call to Action

AI is undeniably a powerful tool, but its true impact depends entirely on how thoughtfully and responsibly we choose to use it. Like any tool, it is neither inherently good nor bad; its effectiveness lies in the hands of those who wield it. Skill, judgment, and a deep understanding of its context are essential to unlock AI's full potential while mitigating its risks. As AI becomes increasingly integrated into our personal and professional lives, we must approach it with a blend of curiosity to explore its possibilities and responsibility to ensure its benefits are shared equitably.

Ultimately, AI is another tool in humanity's ever-expanding toolbox—one that holds the potential to transform how we think, innovate, and interact with the world. By using it wisely, we can shape a future that is not only more efficient but also deeply human, grounded in empathy, ingenuity, and shared purpose. It's a tool that, when used with care, can help us build a better tomorrow.

The responsibility to harness AI begins with each of us. Whether you're an individual exploring how AI can simplify your day-to-day life or a professional integrating AI into your workflow, your actions and approach play a vital role in determining its impact. Learn about the AI tools you use, question how they operate, and advocate for transparency and ethical practices. Most importantly, approach AI as a collaborator, not a solution unto itself. Try it out today. Begin by identifying a routine task AI can assist with, try a tool that addresses it, and reflect on how it improves your workflow or saves you time. Small steps lead to big changes. Now is the time to engage, educate, and innovate with purpose because the future of how we use AI depends on how we start using it today.

Note: this chapter was prepared by the author in his personal capacity. The views and opinions expressed in this article are those of the author and do not necessarily reflect the official policy, opinion, or position of their employer.

About the Author

Brandon Lester is an engineer and technologist with over 20 years of experience in information technology and engineering. Brandon founded Olomana Tech to deliver impactful, innovative solutions for complex challenges across government defense and business sectors. Through Brandon's career he's served as a technical leader at Booz Allen Hamilton (director of AI), Microsoft (sr. technology strategist), SRC Technologies (CTO, cybersecurity engineer), CACI (data engineer), and Northrop Grumman (manufacturing engineer) demonstrating core competencies in cloud technologies, cyber/security engineering, project/program management, data science/engineering, systems engineering, and big data analytics and dashboards. Brandon is also a board director of AFCEA International, previous president of AFCEA Hawaii, and the AFCEA Asia regional vice president, where he helps build communities that connect the defense industry in support of national security. Brandon lives in Hawaii with his wife and two sons.

Email: brandon@olomana.tech
Website: www.olomana.tech

CHAPTER 15

AI: CAN'T BEAT IT, SO JOIN IT

By Peter Lundgreen
Founding CEO, Lundgreen's Capital
Copenhagen, Denmark

The sky is not the limit. Your mind is.
—Marilyn Monroe

When I think of artificial intelligence, the above quote comes to mind, and the reason why I associate AI with Marilyn Monroe's thought is because AI can take our thinking and imagination beyond our own limitations, which is no less than fantastic. The magnitude of the developments in AI has the power to be a game changer, and there's a good chance that AI will change a lot in our own lives. Every time the world changes, some get squeezed and some win. The good news about AI is that everyone has a very good chance to flip the situation towards their favour and avoid being thwarted by the development.

To do so, it will require an individual to spend time understanding several aspects of AI. For example, in terms of one's career, what threats and opportunities could AI create? Doing nothing is already a risk on its own, and the same applies when it comes to AI development in the coming decade, as I expect AI to become more mainstream. This development may eat up jobs and household income, but it's better to join the trend actively since no one can beat it.

As an advocate for technological development, it's easy to come up with an example where an innovation resonates with you deeply. In the United Kingdom, neuroscientists have figured out how to avoid gruelling surgery on children with brain tumours. The key is AI helping analyse a much bigger scope and quantity of data depicting the tumour, which wasn't possible in the past. This helps give a much better understanding of the characteristics of the tumour and helps improve the individual treatment for each child. This is an example of a 100% win. There are already truckloads of new AI developments and technologies that are remarkable. Still, I sincerely hope that AI will lead to new solutions that aren't included in today's or tomorrow's expectations, or maybe even expectations in the future—real game changers.

A fascinating historical example of a game changer, although not AI-related, is the "Great Horse Manure Crisis of 1894". In London and other big cities like New York, horses were used as the engine for transporting people and goods in the 1890s. In London alone, around 50,000 horses were on the streets every day, leaving piles of manure all over the city.

A famous article from The Times in 1894 forecasted that in 1944 the horse manure could reach up to nine feet high in the whole of London. The crisis was even discussed at the international urban planning conference in New York in 1898. Nobody could come up with proposals to solve the problem, and some really saw it as the end of urbanisation. Luckily, technological evolution was able to come up with a solution— the car. By 1912, transportation in big cities was motorised. What makes me wonder slightly about this story is that in around 1890, Gottlieb Daimler already presented his first version of a car, but the openness towards new technology was not of enough magnitude to consider it as a solution.

Today, 99% of all climate researchers tell us that humankind will experience various climate challenges if we don't do anything about it. The soil under our feet could start burning in 2050. I'm a true believer in technological evolution since thousands of years of progress have proven that new technologies solve challenges that initially seemed overwhelming.

Will AI be a substantial part of a solution for the troubled climate? I don't know, but I'm certain that some of the technology-driven rescue solutions for the climate haven't been thought of yet, and they will have the same impact as the car had in 1900. At minimum, AI will influence our lives in small and large matters in the coming decades.

The Cheerful Business Sector

I consider businesses as the true winners of AI. It is a multidimensional win in various aspects of a business. AI is not new, but it is currently growing rapidly, especially after the release of ChatGPT. New technology is always good for business, though I consider AI to have a special positive impact as it opens up for an overflow of new business opportunities.

I also observe the appetite in Lundgreen's, the investment company I founded. Today, Lundgreen's is very active on social media, and we are commercially utilising this part of the world as much as we can. But personally, you won't find me on any social media platform apart from my profile on LinkedIn. Despite the fact that I'm personally not a fan of social media, I'm still fascinated by how our company could commercially benefit from the different social media platforms.

It feels like AI-based initiatives are the natural next moves, and our company will leverage on AI solutions commercially as much as possible, under conditions where it functionally and economically makes sense. The first step where we have invested money in is a specially programmed AI sales assistant that is available on our homepage. This AI sales assistant comes with the linguistic ability to service visitors like an advanced chatbot and can also engage in a conversation to upsell a product or service.

It's an example of how we commercially make use of AI, but internally, I plan to expand AI's role further. In addition to the AI sales assistant on our homepage, we have access to a platform that allows us to program additional AI assistants. Just as personal assistants were once a standard, I envision a future where Lundgreen's Capital employees have their own AI assistants to optimize client service.

A common question about a company is "How many employees does it have?", most likely implying that having more staff is better. If this should be the preferred way of thinking, then I disagree. In the future, I expect our company to answer this question by highlighting both our human workforce and AI assistants specifically programmed for our needs. In the future, if smaller and mid-sized companies can't integrate AI-based solutions in their businesses, then I expect that they are at risk of falling behind.

If I go out of my backyard and zoom in on the giant companies worldwide, I expect to see these giants become even bigger as they benefit the most from the AI trend. I expect them to profit the most, for example, from efficiency gains that the new development in AI functionalities is leading to. The giants are also likely to generate the biggest profits from the latest AI products.

My primary expectation to an even more giant-dominated society is a world with more mass production and mass consumption. More solutions and products will be sourced from an AI base, which is the price to pay for higher efficiency, though more products will probably be produced at lower-quality standards.

Higher efficiency could create hopes for lower prices for consumers, but I doubt this would be the case. Instead, I expect efficiency gains to protect or increase the profit margins of the giants. Developments like efficiency, mass production, and mainstream use sound boring to my ears, though it's the reality that I expect the world to face. Everything has a price.

Intensified Megatrends

In Lundgreen's, we incessantly work with a number of defined mega-trends which we deem to have an impact on the global economy and the financial markets years ahead. One could say that AI is a megatrend itself—and it is, though it also affects other megatrends.

During the past two decades, it was obvious how the incomes of middle-class households in Western economies have been under pressure. In the US, it's to a noticeable extent that these households have been dropping out of the middle-income class and falling into lower brackets with reduced incomes. Normally, healthy and growing economies allow households to move away from poverty and towards the middle-income category. Now, it's in reverse.

Instead, the upward trajectory is observed in many emerging economies. Here, low-income households really move beyond the poverty threshold and approach the lower middle-income and middle-income segment as economies grow at a rapid pace.

I argue that AI will contribute to this trend and could even intensify the development among emerging markets. It is another globalisation driver, despite some politicians thinking they have the power to stop globalisation—when actually, they cannot. On the contrary, the ultimate consequence of this megatrend is that a new global middle class is emerging. Gross household incomes will equalise faster than what we have ever seen, which means that the middle-income consumer base in emerging markets will continue to grow while in the advanced or old economies, the middle-income earners will increasingly feel the pinch.

By breaking language barriers, AI makes it possible to substitute workers from high-cost areas sourced from a global job pool. AI-based tools now translate so accurately that the communication gap has considerably diminished, allowing job vacancies to open up to a wider global labour pool. Employers may choose to replace native speakers of a language by hiring someone from a country with lower labour costs who can effectively utilise AI translators.

There are lots of examples on how well-paid jobs in developed economies are suddenly under threat from the global job market. A few years ago, it was fairly expensive to make a layout for an ad. Today, one can toss all the materials (photo, headline, text, etc.) into the AI pot and out comes several design suggestions within a few seconds. If the advertiser wants something more upscale, there are qualified freelancers in the global job market that can handle the assignment at a competitive rate. Everybody in that business is now using AI-based tools for good quality at a low price—and that is another example of how AI leads to the globalisation of the labour market and the equalisation of salaries.

The AI Challenge for Individuals

Do I think that AI is, or will do, something good for me? Certainly. Many things have become quicker and easier nowadays, and I know that AI is the reason behind these positive improvements. Thinking a bit broader and bigger, AI will then help improve medical treatments and healthcare even more significantly, and aside from this, I'm sure that more people will point at a range of other areas and qualities where AI could help, maybe even social opportunities.

Another aspect for individuals that I'm pondering on is personal security. It's the price we're expected to pay for AI "progress" across the world. We have already seen how the challenge with fake photos exploded, and it will continue with the replication of voices, handwriting, eye irises, and even full identities.

To use my own example, I only appear in the media as a professional person, normally not as the private person Peter Lundgreen. Still, it's natural for me to mention some personal information during an interview, podcast, TV guest appearances, etc. For almost a decade now, I have had several hundred media appearances per year. With AI technology, all these materials can be filtered and run together with all kinds of registers and other databases about where I have lived, worked, studied, etc. My expectation is that it's possible to make a full copy of my personal identity and also come up with a list of the 25 most likely passwords I would choose based on all the nuggets of information publicly available.

I'm not afraid, as we can't stop living life either, but I acknowledge the risk, and I expect to purchase better cyber protection products and insurances. I anticipate that we will all be more vulnerable to numerous attacks on all our accounts, and we will need to buy protection just to continue using a smartphone, and some people might even be forced to leave social media platforms. I'm not convinced that governments will be able to guarantee personal electronic AI safety, so at a minimum, an individual person will have to invest in a more secure level of protection, though it will not come cheap.

The individual risk is one of my concerns, but let's leave this potential risk behind for a moment and let cyber dreams fly instead. AI has a family member, "metaverse," the technology that is presently used to optimize gaming experience. In this part of the AI world, I argue that there are fascinating options for users.

Alchemy has been a dream for centuries, or put in another way, the search for a fast track to wealth or at least finding a more comfortable life with less work. In some countries or business segments, the trend after the Covid-19 period has been to upgrade the private part of life and downscale work life—a half-baked solution towards the idea of lying down in a hammock at a sunny beach while the monthly salary is credited on the account. Alchemy has never worked, and currently, the hammock dream is also taking a headwind as staff members are increasingly called back into the office.

If the Web3 world including AI were to be fully developed, it has the potential to change our work life fundamentally. We will finally have a fully digital metaverse version of ourselves, which will be practical in our private lives. The digital avatar will also be the perfect representative of an employee in the virtual version of an office. Here, creativity can be utilised, like in physical meetings, something a regular virtual meeting doesn't offer, so the virtual avatar office might be the best alternative to be out of the office forever.

The ultimate metaverse world with a digital avatar copy of all of us implies only one challenge: the energy consumption to run such a digital world is bigger than anyone could imagine and would cause climate crises number two and three. Yes, alchemy with access to endless wealth without working has so far been impossible, and it might be that even AI cannot fulfil that dream after all.

About the Author

Peter Lundgreen has the superpower to make topics about finance and the global markets easy to digest and thought-provoking. With four decades of experience in the finance industry and as founding CEO of his own successful investment house, Peter shares deep insights about developments in the global markets through articles and podcasts, international TV and radio appearances, and speaking at major industry events around the world.

Peter is an advocate of global start-up companies that are shaking things up in technology, defence, and green energy—the focus of his remunerative alternative investment funds. He also takes active roles in mentoring and advising entrepreneurs on how they can successfully develop their businesses.

During his spare time, Peter enjoys being in the kitchen, scouting the world for the best wines, and listening to ABBA.

Website: www.lundgreensinvestorinsights.com

CHAPTER 16

EMPOWERING GEN Z THROUGH AI-DRIVEN COACHING AND MENTORSHIP: A PATH TO SUSTAINABLE GROWTH

By Veejay Madhavan
Founder, OulbyZ; Multigenerational Workforce Strategist
Singapore, Singapore

> *The future of artificial intelligence is not about man versus machine, but rather man with machine. Together, we can achieve unimaginable heights of innovation and progress.*
> —Fei Fei Li

In an era marked by rapid change and technological innovation, Generation Z is emerging as a transformative force within the workforce. Born into a digital world, Gen Z individuals are not only tech-savvy but also highly adaptable, with a strong appetite for continuous feedback and personal growth. Their expectations for instantaneous communication and enhanced engagement with leadership are challenging and reshaping traditional professional development models.

Empowering Gen Z through effective coaching and mentorship isn't merely advantageous—it's essential. Their openness to guidance, combined with their ease with technology, offers a unique opportunity to harness AI-driven approaches. In this chapter, I explore how integrating artificial intelligence into coaching and mentorship can meet Gen Z's distinctive needs, foster sustainable growth, and pave the way for a more innovative and resilient future.

Distinctions Between Coaching and Mentoring

Understanding the differences between coaching and mentoring is crucial for effectively empowering Gen Z. While both aim to support personal and professional development, they diverge in terms of initiation, responsibility, and the dynamics between the coachee and the coach or the mentor and the mentee.

Coaching is a collaborative process initiated by the coachee, who identifies their own goals and steers their developmental journey. The coachee assumes full responsibility for their progress and outcomes—whether successes or setbacks. The coach serves as a facilitator, aiding the coachee in self-discovery and decision-making without providing direct solutions. This approach focuses on self-driven growth, where individuals proactively set personal objectives and seek improvement. Coaches use probing questions and reflective tools to help unlock potential, emphasizing empowerment through insight and fostering self-reliance.

For example, consider a Gen Z employee aiming to boost their leadership abilities. They seek out a coach, and through thoughtful questioning, the coach helps them reflect on their strengths and areas for improvement, leading to an actionable development plan. The coachee takes ownership of executing this plan and monitoring their progress, fully embracing personal accountability.

Mentoring, on the other hand, involves the mentor taking an active role in guiding the mentee, sharing knowledge, experiences, and advice. This relationship often entails a shared investment in outcomes, with the mentor feeling a sense of responsibility for the mentee's development. Mentoring is characterized by mentor-led guidance, where mentors identify growth opportunities and suggest developmental paths. They provide insights and answers based on experience, developing a relationship over time built on trust and mutual respect. An example of this would be an experienced professional mentoring a Gen Z newcomer, offering insights into industry trends, career advancement strategies, and facilitating networking opportunities. Both mentor and mentee are invested in the mentee's success, emphasizing shared responsibility and motivation through connection.

Gen Z's values align closely with the coaching model due to their desire for autonomy, self-direction, and accountability. They prefer to be in charge of their growth, seeking facilitators who empower them rather than dictate terms. However, I recognize that mentoring remains valuable by providing access to seasoned wisdom and guidance in navigating uncharted territories. By discerning when to apply coaching versus mentoring and leveraging their unique benefits, I believe organizations can design development programs that resonate with Gen Z's preferences.

Generational Characteristics: Digital Natives Shaping the Future

Gen Z's upbringing in a digitally immersed environment significantly influences their expectations and interactions. Their proficiency with technology, preference for instant information access, and emphasis on authenticity shape how they engage with coaching and mentorship. They seamlessly integrate digital tools into their daily lives, seeking alignment between their work and personal values—a purpose-driven outlook. Their desire for immediate and ongoing feedback guides their progress, and they favor teamwork and collective problem-solving approaches. As Simon Sinek insightfully remarks, "People don't buy what you do; they buy why you do it." This perspective aligns with Gen Z's focus on purposeful engagement, highlighting the importance of aligning coaching and mentorship initiatives with their core values.

When to Apply Coaching

Coaching is particularly effective for Gen Z in scenarios where immediate feedback and rapid skill development are essential. Situations where coaching is most beneficial include skill development—enhancing specific abilities like communication, leadership, or technical expertise. It is also valuable in performance optimization, tackling challenges that impact current job effectiveness, and in career navigation, assisting with transitions into new roles or responsibilities. Moreover, coaching supports goal achievement by helping set and pursue short-term professional objectives.

Coaching resonates with Gen Z due to their desire for swift progress, aligning with their preference for quick results and tangible outcomes. The customization inherent in coaching tailors the experience to individual goals and learning styles, encouraging empowerment by taking ownership of their personal and professional growth.

AI's Role in Facilitating Coaching and Mentoring

The integration of artificial intelligence (AI) into coaching and mentoring introduces new dimensions of personalization, accessibility, and efficiency, perfectly suiting Gen Z's expectations. AI-driven coaching unlocks sustainable growth by offering personalized development journeys. AI algorithms tailor development plans based on the individual's strengths, weaknesses, and preferences. An AI coach might evaluate a coachee's skill set and recommend targeted activities to enhance specific competencies. Immediate feedback from AI provides real-time insights, enabling swift adjustments and continuous improvement—such as offering instant analysis of a sales pitch and highlighting areas for refinement.

Accessibility and scalability are enhanced through AI's around-the-clock availability. AI coaches are accessible anytime, accommodating flexible schedules. A coachee might engage in skill-building exercises during off-hours, supported by AI-driven guidance. Cost-effective solutions reduce financial barriers, making professional development more widely available.

AI enhances engagement through interactive experiences. Gamified elements and interactive modules make learning engaging and enjoyable; for example, a coachee earns badges for completing milestones, fostering motivation. Advanced AI can detect emotional cues and respond empathetically, adding an element of emotional intelligence to the coaching process.

AI also plays a significant role in enhancing mentoring by building meaningful connections. Intelligent matching through AI analyzes profiles to match mentees with mentors who align with their goals and interests. A mentee passionate about environmental sustainability might be paired with a mentor experienced in green technologies. AI bridges geographical gaps, facilitating connections across the globe and enriching mentorship with diverse perspectives. For instance, a mentee in Asia connects with a mentor in Europe, gaining international insights. Personalized resources are provided through AI's curated content, recommending articles, videos, and learning materials relevant to mentorship discussions. Streamlined communication is achieved with AI's assistance in scheduling, reminders, and follow-ups, enhancing the mentoring experience.

Critical Knowledge and Skills for AI-Driven Coaching and Mentorship

Effectively leveraging AI in coaching and mentorship requires a blend of technical understanding and interpersonal skills. Digital literacy is essential—understanding AI fundamentals enhances the ability to utilize technology effectively, while data interpretation skills inform more personalized and impactful guidance.

Ethical awareness is crucial. Commitment to ethical practices ensures responsible use of AI, and bias mitigation involves recognizing and addressing potential biases in AI systems, promoting fairness. Emotional intelligence is vital to maintain human connection; balancing technological efficiency with empathy ensures meaningful interactions. Cultural sensitivity enhances inclusivity, respecting diverse backgrounds.

Adaptability plays a significant role. Embracing innovation involves openness to new methodologies, fostering continuous improvement. Responsive learning incorporates feedback from AI insights, refining coaching and mentoring approaches.

Explainable AI in Coaching and Mentoring

As AI becomes more integrated, recognizing the need for transparency—known as explainable AI—is vital. Building trust through transparency enhances user confidence, as clear explanations of AI decisions empower users. Understanding AI recommendations enables informed decision-making, fostering user empowerment.

Explainable AI enhances ethical practices by helping identify and correct biases, promoting ethical standards. Transparency meets legal requirements for automated decision-making, ensuring compliance. Increased effectiveness is achieved through personalized explanations, strengthening the relevance of AI recommendations. Collaborative interaction allows users to provide feedback, refining AI performance over time.

Ethical Considerations in Using AI

Ethical considerations are essential to ensure AI is used responsibly in coaching and mentoring. Protecting personal information with robust measures safeguards sensitive data. Transparent policies and clear communication about data use build trust. Fairness and inclusivity are promoted by preventing bias; regular reviews ensure algorithms are equitable, and inclusive datasets and team diversity reduce biases.

Preserving the human element is crucial. A balanced approach that combines AI efficiency with human empathy maintains meaningful connections. Customization by adjusting AI interactions to suit individual preferences enhances the experience. Accountability involves clear delineation of human oversight to ensure ethical standards are upheld.

Translating AI Integration to Sustainable Growth

Incorporating AI into coaching and mentoring can drive sustainable growth on multiple levels. Individual empowerment is achieved through continuous learning—personalized, adaptive learning fosters lifelong development, equipping individuals to navigate an evolving professional landscape with resilience.

Organizational advancement is facilitated through enhanced performance; tailored development boosts productivity and innovation. Meeting Gen Z's expectations improves satisfaction and loyalty, aiding talent retention. Societal impact is realized through collective progress, as empowered individuals contribute positively to society. Promoting responsible AI use sets a precedent for ethical leadership and future practices.

Conclusion

The synergy between Generation Z's pioneering spirit and the vast possibilities of artificial intelligence offers unprecedented opportunities for growth and innovation. By thoughtfully integrating AI into coaching and mentorship, tailored to the unique characteristics and preferences of Gen Z, we can unlock individual potential and foster sustainable advancement for organizations and society alike.

Looking ahead, embracing AI in coaching and mentorship is more than a strategic advantage—it's a necessary evolution. It's an invitation to harness technology responsibly, cultivate innovation, and build a resilient foundation for the future.

About the Author

Veejay Madhavan is a business strategist with over 26 years of experience in people strategy, business transformation, and leadership development across Asia. As the founder of OulbyZ, he bridges the generational gap in workplaces and helps organizations leverage AI to build sustainable high-performing, multi-generational teams. His work extends to educational institutions, where he pioneers AI-driven simulation labs to bridge the skills gap between Gen Z and real-world corporate challenges. Currently pursuing a doctoral degree at Golden Gate University, his research focuses on the impact of Generative AI on workforce dynamics.

Veejay is the co-author of an upcoming book on AI's impact on Human Capital Management. His contributions to Gen Z learning and workforce innovation have earned him accolades, including The Visionaries Award 2024 and the Outstanding Leadership Award. OulbyZ was also recognized in Silicon India's '20 Most Promising Companies in APAC'.

An angel investor and mentor to startup CEOs, Veejay also supports his Gen Z daughter's dream of a golfing career.

Email: askus@oulbyz.com
Website: www.oulbyz.com

CHAPTER 17

LEVERAGING DATA ANALYTICS AND AI FOR REAL-TIME DECISION-MAKING IN FINANCE AND BEYOND

By Ghofran Massaoudi
AI Expert, Data Scientist, Consultant
Tunis, Tunisia

Information is not knowledge. The only source of knowledge is experience. You need experience to gain wisdom.
—Albert Einstein

Introduction: The Age of Data-Driven Decisions

We live in a data-driven era where organizations, particularly in finance, increasingly rely on data analytics and artificial intelligence (AI) for decision-making and growth. In a volatile and complex financial sector, real-time data analysis is revolutionizing decision-making processes. Coupled with AI, data analytics enables automation, optimization, and innovation that surpass human capabilities. This chapter examines how the integration of AI and real-time analytics is transforming finance and setting standards for other industries.

Predictive and Prescriptive Analytics

At the heart of data-driven decision-making lies predictive and prescriptive analytics. Predictive analytics leverages historical data to forecast future outcomes, enabling financial institutions to anticipate risks, market fluctuations, and customer behaviors. By employing machine learning models to analyze patterns within extensive datasets, these institutions can generate highly accurate predictions. For instance, they can forecast stock market trends, assess creditworthiness, and identify potential fraud before it occurs.

Prescriptive analytics takes this a step further by providing specific recommendations based on the predictions made. For example, it can guide a financial institution in determining optimal strategies for investment portfolio enhancement or risk management. With the help of AI, these systems can continuously learn from new data, thereby enhancing the precision of their recommendations over time.

Real-Time Data Processing

In the financial sector, timing is critical. The capability to process vast amounts of data in real-time empowers organizations to make quick decisions that can have profound financial consequences. High-frequency trading, for instance, involves executing stock trades at speeds that far exceed human capabilities. AI algorithms continuously monitor financial markets, executing trades based on predefined criteria.

Beyond trading, real-time data processing is essential for effective risk management. Financial institutions can now instantaneously track global events, economic indicators, and even social media sentiment to evaluate risk levels. This proactive strategy enables organizations to address potential risks before they escalate into crises.

The Evolution of AI in Finance

Automating Traditional Processes

AI is transforming many traditional financial processes by automating routine tasks, allowing human professionals to concentrate on more strategic decision-making. For example, AI can streamline loan approval processes by efficiently analyzing credit histories and associated risk factors, surpassing the effectiveness of conventional methods. This automation not only accelerates the process but also minimizes human error and bias.

Fraud Detection and Cybersecurity
One of the most crucial applications of AI in finance is in fraud detection and cybersecurity. As cyber threats grow more sophisticated, financial institutions must proactively defend against potential attacks. AI-driven systems monitor transactions in real time, detecting suspicious activities based on recognized patterns and anomalies. These systems can flag unusual behaviors, such as large withdrawals from unfamiliar locations or numerous failed login attempts, alerting security teams before any damage occurs.

Moreover, AI is essential in bolstering cybersecurity measures by identifying new and evolving threats, learning from past incidents, and adjusting defenses accordingly. This real-time threat analysis is vital in an increasingly digital landscape where cybercriminals are constantly adapting their tactics.

Expanding AI's Reach Beyond Finance

AI and data analytics have already transformed the finance industry by enabling organizations to make real-time decisions, detect fraud, manage risks, and optimize portfolios. As technology progresses, its impact is broadening across various sectors, reshaping how businesses operate and compete. This section explores specific applications and emerging trends in finance, healthcare, retail, transportation, and beyond.

AI for Anti-Money Laundering (AML)

The battle against money laundering has long been a critical focus for financial institutions, yet it remains a complex challenge. AI is emerging as a powerful ally in anti-money laundering (AML) efforts by automating the detection of suspicious activities across transactions. These AI systems can analyze vast amounts of financial data in real time, identifying unusual patterns and anomalies that may signal money laundering attempts.

For instance, AI can detect sudden surges in transaction volumes, transfers to high-risk countries, or patterns of repeated small deposits followed by large withdrawals—behaviors often linked to illicit activities. By reducing false positives, which have plagued traditional AML processes, these AI-driven systems allow compliance teams to concentrate on high-risk cases more effectively.

AI in Other Industries: Expanding Horizons

While AI's transformative impact on the financial sector is well-established, its influence is rapidly spreading to other industries. Below are examples of how AI and data analytics are reshaping sectors such as healthcare, retail, and logistics.

Healthcare: Real-Time Diagnostics and Treatment

AI is increasingly being integrated into healthcare systems to provide real-time diagnostics, personalized treatment plans, and predictive health monitoring. The emerging field of "precision medicine" heavily relies on AI's capability to analyze genetic data, medical records, and wearable device data to tailor treatments for individual patients.

In healthcare, AI's ability to process data in real time is revolutionizing diagnostics and treatment planning. For example, AI algorithms can evaluate medical images instantly, quickly identifying anomalies like tumors or other critical health issues, often faster than human radiologists.

Telemedicine and Virtual Care

AI has also catalyzed the growth of telemedicine and virtual care, especially in response to the COVID-19 pandemic. AI-powered platforms can triage patients based on their symptoms, provide medical advice, and schedule appointments, thereby streamlining the patient experience and alleviating the workload on healthcare providers.

Furthermore, AI facilitates personalized medicine by analyzing genetic information, lifestyle data, and medical histories to recommend tailored treatment plans. This not only enhances diagnostic accuracy but also allows for real-time health monitoring, enabling physicians to adjust treatment strategies as patient conditions change.

AI in Retail: Enhancing Customer Experience

AI is significantly improving customer experience in the financial sector. AI-powered chatbots and virtual assistants have become ubiquitous in banking, offering round-the-clock customer support. Utilizing natural language processing (NLP), these tools can understand and respond to customer inquiries in real-time, providing personalized financial advice and assistance. Additionally, by analyzing customer behaviors and preferences, AI enables banks to customize their services, delivering more relevant products and ultimately enhancing customer satisfaction.

Recommendation Engines

In e-commerce, AI-driven recommendation engines analyze browsing behavior, purchase history, and social media interactions to deliver personalized product recommendations in real time. This approach not only boosts customer satisfaction but also increases sales by offering a more targeted shopping experience.

Retail giants like Amazon have perfected this strategy, using AI to suggest complementary products, predict future purchases, and even optimize pricing strategies based on demand. Additionally, AI helps forecast consumer behavior during promotional events, such as Black Friday or holiday sales, enabling companies to fine-tune their marketing efforts and manage inventory effectively.

AI in Credit Risk Management

AI is revolutionizing how financial institutions assess credit risk. Traditionally, credit scoring systems relied on a limited set of data points, such as income and credit history, to determine an individual's creditworthiness. Today, AI-powered systems can analyze vast datasets, incorporating alternative data sources like social media activity, online behavior, and geolocation to create a more comprehensive profile of borrowers.

By utilizing machine learning algorithms, AI can uncover subtle patterns and signals that human analysts might overlook. This capability enables financial institutions to make more accurate predictions regarding borrowers' abilities to repay loans, enhancing risk management practices. Moreover, AI streamlines the credit assessment process, reducing the time and cost associated with manual evaluations. As a result, lenders can expedite loan approvals and extend credit to a broader range of customers, including those lacking traditional credit histories.

Retail: Enhancing Supply Chain Management

The retail sector is also harnessing AI for real-time decision-making. AI-driven recommendation engines analyze customer behavior instantaneously, providing personalized product suggestions that enrich the shopping experience. Retailers utilize AI to assess trends, customer feedback, and social media insights to refine their marketing strategies, pricing models, and inventory management.

In addition, AI and data analytics are significantly enhancing supply chain management. By monitoring supply and demand in real-time, AI assists retailers in adjusting their inventories and logistics processes, resulting in cost reductions and improved efficiency. This capability allows companies to anticipate disruptions in the supply chain—whether from natural disasters, political events, or market fluctuations—and respond proactively.

Emerging Trends in AI-Driven Decision-Making

As AI and data analytics technologies continue to advance, several key trends are emerging that will influence the future of real-time decision-making across various industries.

AI in Autonomous Finance: Robo-Advisors
Robo-advisors are AI-powered platforms that offer automated financial advice and portfolio management services. These platforms leverage algorithms to evaluate an investor's risk tolerance, financial objectives, and market conditions, delivering personalized investment strategies without the need for human intervention.

The emergence of robo-advisors has democratized access to financial advice, enabling individuals with smaller portfolios to benefit from professional-grade investment management at a fraction of the cost. Furthermore, these platforms continuously monitor market fluctuations, adjusting portfolios in real time to maximize returns or mitigate risks.

Transportation and Logistics: Optimizing Routes and Reducing Costs
In transportation and logistics, AI is enhancing operations by optimizing routes in real time. Companies like Uber and Amazon employ AI algorithms to anticipate traffic patterns, weather conditions, and fuel costs, ensuring timely deliveries. AI also aids fleet management by analyzing sensor data from vehicles to predict maintenance needs, thereby minimizing downtime. For instance, predictive analytics can forecast when a vehicle requires maintenance based on historical data, allowing for proactive repairs and avoiding delivery delays.

The Integration of AI with Blockchain for Enhanced Security
AI and blockchain technologies are converging to improve security and transparency in sectors requiring secure data transactions, such as finance and healthcare. While blockchain provides a decentralized and immutable transaction ledger, AI enhances the efficiency and security of these transactions. In finance, this combination boosts fraud detection and secures transactions, as AI analyzes blockchain data in real time to identify threats and prevent unauthorized access, reducing cyberattack risks.

Data Privacy and Security
As organizations collect more data, privacy and security concerns escalate. Industries like finance and healthcare must ensure their AI systems comply with data protection regulations such as GDPR, implementing strong encryption and secure storage practices. AI systems must also maintain transparency in decision-making processes, especially in finance, where opaque algorithms could undermine trust if customers and regulators can't understand how decisions—like loan approvals—are made.

The Impact on Employment
AI's role in automating tasks raises concerns about its impact on employment. While it improves decision-making and operational efficiency, it also threatens jobs traditionally held by humans, such as traders and analysts in finance due to automated trading systems. However, AI also generates new opportunities; demand for professionals skilled in AI development, data science, and ethics will grow. Moreover, humans will remain vital in overseeing AI systems, ensuring accuracy, and solving complex problems that require human judgment.

Bias in AI
AI systems reflect the data they are trained on, meaning biases in that data can lead to discriminatory practices in finance, such as biased lending or investment recommendations. It's essential to train AI on diverse, representative datasets to prevent these biases and ensure fairness in decision-making.

Ethical and Regulatory Considerations
As AI usage increases, so do ethical and regulatory concerns surrounding data privacy, algorithm transparency, and bias. Ensuring fairness and

transparency in AI decision-making is crucial, especially in critical areas like hiring and lending. Regulations like GDPR push organizations to adopt ethical AI practices that protect customer data.

The Future of AI and Employment
While AI enhances efficiency and creates new opportunities, it may also lead to job displacement in various sectors, including finance and retail. However, it will generate new roles in data science, AI development, and ethics, necessitating reskilling and upskilling for workers to remain competitive.

Conclusion: A New Era of Decision-Making

AI and data analytics are transforming real-time decision-making across industries, enabling faster, informed decisions. This evolution is also reshaping healthcare, retail, and logistics by improving efficiency and customer experiences. Yet, challenges like data privacy, employment impacts, and bias must be addressed. Organizations that effectively leverage AI will be best positioned to thrive in this data-driven future.

References

Real-Time Data Processing in Finance

- Deloitte Insights: "How Artificial Intelligence Is Reshaping the Financial Services Industry"

- McKinsey & Company: "Artificial Intelligence in Finance: The Road Ahead"

AI in Enhancing Customer Experience

- PwC: "Real-Time Data Analytics: A Game-Changer in the Financial Services Industry"

- Accenture: "How Artificial Intelligence is Used in Fraud Detection"

Fraud Detection and Cybersecurity

- IBM: "AI and Fraud Detection in Finance: A Growing Necessity"

- McKinsey & Company: "Artificial Intelligence in Finance: The Road Ahead"

AI for Anti-Money Laundering (AML)

- Deloitte Insights: "How Artificial Intelligence Is Reshaping the Financial Services Industry"

AI in Other Industries: Healthcare and Retail

- Harvard Business Review: "How AI is Changing Healthcare"

- World Health Organization: "Artificial Intelligence in Healthcare: Opportunities and Challenges"

- Capgemini Research Institute: "AI in Retail: The Dawn of a New Era"

AI in Credit Risk Management

- McKinsey & Company: "Artificial Intelligence in Finance: The Road Ahead"

Emerging Trends in AI-Driven Decision-Making

- Gartner: "AI-Driven Supply Chain Optimization"

Bias in AI and Ethical Considerations

- Brookings Institute: "AI Ethics: Balancing Innovation and Risk"

- Future of Privacy Forum: "The Role of Ethics in AI and Data Analytics"

The Future of AI and Employment

- Deloitte Insights: "How Artificial Intelligence Is Reshaping the Financial Services Industry"

Conclusion: A New Era of Decision-Making

- Various sources including the above-mentioned ones for a comprehensive view.

About the Author

Ghofran Massaoudi is an emerging professional in the fields of data science and artificial intelligence, currently pursuing a master's degree in intelligent decision-making strategies. With a diverse background that spans roles in human resources, data analytics, and software development, Ghofran brings a unique perspective to the integration of technology in decision-making processes.

Passionate about leveraging AI to enhance organizational effectiveness, Ghofran has contributed to various projects that focus on automating and optimizing workflows, especially in the financial sector. As a dedicated volunteer and consultant, Ghofran is committed to fostering educational initiatives that empower high school students with essential skills in technology and human resources.

With a keen interest in the ethical implications of AI and data usage, Ghofran strives to promote transparency and fairness in technology applications. This blend of technical expertise and a human-centric approach positions Ghofran as a thought leader in shaping the future of AI-driven decision-making.

Email: massaoudi.ghofran@gmail.com

CHAPTER 18

A CALL TO DEFEND HUMANITY

By Grzegorz P. Mika
AI and Cognitive Scientist, Entrepreneur
Chybie, Poland

> *The human spirit must prevail over technology.*
> —Albert Einstein

Humanity has long been fascinated—and haunted—by visions of its own demise. From ancient prophecies to modern science fiction, we've spun countless narratives around the idea of our own annihilation. In recent years, as scientific advancements accelerate and artificial intelligence takes giant leaps forward, these visions seem to grow eerily closer. Compare the world today to just five years ago: has the looming danger become more tangible than we imagined? And if it is, what form does this threat take, and can it be stopped?

As a PhD student in artificial intelligence, with over five years dedicated to unraveling the complexities of advanced machine learning algorithms, I've found myself at the intersection of two deeply held values. On one side stands the pursuit of scientific progress for the benefit of humanity; on the other, the timeless principles of truth, goodness, and wisdom that lie at the core of human existence. For me, it feels as though the line has already been crossed. Yet I still hold hope that we can strike a balance between advancing AI technology and upholding the essential values that define us as human.

For years, one question has lingered: will there come a moment when artificial intelligence begins to "live" a life of its own, and if so, what will that mean for humanity? Can AI ever be truly aware of its own existence, capable of consciousness or a soul? When it generates text or composes music, is it expressing something from the depths of its being—or is it simply mimicking patterns without any self-awareness? And, ultimately, how do these possibilities intersect with the core values that define us as human?

Here, I invite you on a journey of self-discovery in a world reshaped by artificial intelligence. Together, we will examine the potential threats this new reality poses and explore practical strategies to navigate and overcome them. Along the way, we'll delve into real-world examples that reveal the complex, multifaceted challenges AI presents—challenges that, when understood, can empower us to face the future with confidence and insight.

Civilization as Humanity's Mistake

A technologically advanced civilization can be known by its ability to annihilate itself. At first, you might think this sounds like something straight out of a science fiction movie—an idea confined to dystopian futures where machines rise up against humanity or where we destroy ourselves in a dramatic, cataclysmic event. But what if I told you that this is already happening, although not in the way you might imagine? The threat isn't as obvious as a violent takeover or a global collapse. Instead, it's something much quieter, creeping in unnoticed, and it's affecting you—right now.

Imagine waking up one fine morning to the sound of a bell. You groggily rise from your bed, unsure of what the day will bring. As you stumble toward the door and open it, you are met with the cold, unyielding faces of security forces. They hold up a legal document—a warrant for your arrest. Confusion overwhelms you as you try to comprehend the situation. The officers explain that you've been implicated in a serious crime, but as the interrogation unfolds, a shocking twist emerges: the name you've known all your life—your identity—is not your own. The real you is someone completely different, and this revelation is not a mistake or a misunderstanding.

This startling revelation mirrors the plot of the Hollywood film *The System*, directed by Irwin Winkler in 1995. Nearly 30 years ago, no one could have foreseen that a single moment would make identity swapping a viable reality. The story of identity swapping that once seemed like pure fiction has now become eerily prophetic, and as we continue to navigate a world where our digital lives are intertwined with our physical ones, the lines between fact and fiction are increasingly difficult to discern.

You could now ask, 'What is actually true?' In an age where information flows constantly and unfiltered, the question has never been more pressing. The threat of fake news, and especially the monumental challenge of verifying the truth, has become one of the most significant issues of our daily lives today. With the rise of social media platforms, deepfakes, and the rapid spread of misinformation, distinguishing fact from fiction has become increasingly difficult. Our trust in news sources is eroding, and the boundaries between credible journalism and fabricated stories are often blurred. The consequences are far-reaching—affecting everything from political elections to public health decisions. In this environment, skepticism and critical thinking are necessary tools, but even they can be manipulated. The challenge we face is not only understanding what is true but also learning how to navigate a digital world where the truth is often elusive.

Think about AI in two different places: in physical devices and in the virtual world. The AI in physical devices is still something we can mostly control so far. It acts according to rules we set, and it only makes decisions within the boundaries we give it. But when it comes to AI in the virtual world, it's a different story. This kind of AI seems to be slipping out of our grasp. We're starting to see it evolve in ways that we can't always predict or control, and we might eventually reach a point where it's operating so independently that the only way to stop it could be to shut down the entire internet.

But here's the catch—shutting down the internet would have massive consequences. We're all so connected now, and our identities are deeply intertwined with the digital world. If the internet were to go down, it wouldn't just be a technical glitch—it would be like losing a part of yourself.

This paints a pretty worrying picture, doesn't it? It shows just how far we've come and how quickly things are evolving. Some of the consequences of this rapid technological growth might already be irreversible. It's incredible that we've created all this, but the real challenge isn't just in building the technology—it's in learning how to use it responsibly and keeping it in check.

Technology Is Killing Our Insides

Elon Musk told us, "There is some chance that is above zero that AI will kill us all. I think it's low. But if there's some chance, I think we should also consider the fragility of human civilization." This statement encapsulates the deepening anxiety shared by Musk and a multitude of AI experts. In light of these concerns, the Future of Life Institute has published a compelling letter signed by leading entrepreneurs and AI experts, advocating for a responsible approach to AI development. They emphasize that neglecting the potential dangers of advanced AI could lead to dire consequences for humanity. Even if you share these global concerns, how can you be certain that you're not unwittingly exposing yourself to AI threats from a less informed standpoint?

Of course, increasing our reliance on artificial intelligence undoubtedly makes us more vulnerable to system failures, deepfakes, and privacy violations. However, we must not overlook the more subtle yet significant risks we face daily. A greater dependence on artificial intelligence will inevitably lead to a decline in our human skills and diminish our capacity to solve problems independently. It is crucial to address these challenges head-on.

It could be said that the biggest threat is not AI itself, but rather it is people equipped with the latest technology and the uncontrolled development of artificial intelligence. Despite many efforts from experts, scientists to politicians, we are currently living in a time where the development of artificial intelligence is uncontrolled, potentially leading to many threats, including action against, but not yet independence.

Before long, we may find ourselves becoming more like the machines we've created—disconnected from our feelings, numb to the world around us, and increasingly dependent on technology. Think about your daily life. What once felt like a tool to make things easier has now become something you can't imagine living without. Your phone, your laptop, your smart devices—they're everywhere, always at your fingertips. They connect you to the world, to others, and yet, in doing so, they create a barrier between you and the world that matters most. Look around the next time you're with friends or family. How often do you see a room full of people, sitting together but absorbed in their screens, not speaking a word to each other? The conversations you used to have, the moments of connection, seem to have vanished, replaced by a quiet hum of technology. The irony? These devices, the very things you rely on to bring you closer to others, are also driving a wedge between you.

In this digital age, it's no longer just the machines we should fear. The real threat may lie in humans themselves—humans who have become so intertwined with their devices that we risk losing our humanity in the process. The latest advancements in technology, the very tools that were designed to improve our lives, have instead begun to redefine us. We are becoming, in a sense, extensions of our devices—our thoughts, emotions, and connections reduced to algorithms and notifications. Artificial intelligence may one day surpass human intelligence, but right now, the greatest danger lies in what we allow technology to do to our minds, our relationships, and our ability to simply be present in the world.

The process of evolution is relentless and absolute. When a function in nature is no longer needed, it tends to fade away. As artificial intelligence increasingly takes over tasks like critical thinking and problem-solving, there is a significant risk of irreversible consequences. If our brains are responsible for these cognitive functions, outsourcing them to AI could lead to the atrophy of these mental faculties. Over time, this may result in a decline in our intelligence and resourcefulness. Dr. Umberto Leon Dominguez from the University of Monterrey is among the first researchers to explore this phenomenon, and his findings lend support to these concerns.

Pablo Picasso told us, "To paint, you must close your eyes and sing." They say a picture is worth a thousand words, and painting can be a deeply cathartic process, helping us tap into emotions hidden deep within our subconscious. We've all probably come across images online that seem unbelievable or even controversial, only to later discover they were AI-generated fakes. If you take a closer look at the details, you might notice things that just don't add up. I recently came across a photo of autumn in a New Hampshire town, showcasing its vibrant colors. But there was a catch—the trees were unnaturally bright, and it didn't take long for eagle-eyed users to spot the oddity. Sadly, many were tricked at first, which goes to show just how easily technology can deceive us. It's a clear reminder of how trusting we've become of tech. That's why it's so crucial to stay vigilant, not just for ourselves, but also to help our loved ones and those around us stay informed and skeptical of everything technology puts in front of us.

If AI is capable of generating photos, it stands to reason that it can also create images with even greater precision and accuracy. Rather than getting caught up in the complex issues of copyright laws, the role of AI in the creative economy, and the financial implications, let's leave those to the appropriate authorities. Instead, let's focus on something even more crucial: the profound value that traditional painting and drawing holds for our cognitive and emotional development.

Painting nurtures creativity, promotes emotional well-being, enhances problem-solving abilities, and boosts memory and concentration. These are skills that contribute to a well-rounded and resilient mind. Now, imagine a world where children in schools no longer engage in painting, but instead are taught to generate images with artificial intelligence. What would happen if we stopped fostering these vital qualities and skills? It's difficult to envision such a future, but it's important to consider what we might lose in the process.

Not only painting and drawing contribute to the development of our minds, but so does writing—particularly when it's done by hand. In today's digital age, however, pen and paper are increasingly being replaced by digital devices. As a result, handwritten notes are becoming less common in schools and universities. There are several reasons behind this shift: typing on a keyboard is faster, and digital literacy has become a fundamental skill in the modern job market.

While these technological advancements offer undeniable benefits, it's important to remember that handwriting has unique cognitive advantages. It fosters deeper learning, enhances memory retention, and strengthens fine motor skills in ways that typing and generative AI cannot replicate. As we embrace the convenience of digital tools, we should also be mindful of the importance of maintaining the practice of writing by hand in our educational systems.

Researchers at the Norwegian University of Science and Technology in Trondheim have conducted studies comparing the engagement of neural networks in the brain when writing by hand versus typing on a keyboard. Their findings reveal that when we write by hand, the brain's connection patterns are significantly more complex and widespread than when typing. This heightened brain activity plays a crucial role in forming memory traces and encoding new information, which is essential for effective learning.

Moreover, the act of writing—whether it's essays, books, or poems—goes beyond just the mechanical process of recording words. It stimulates our imagination, helps us express our thoughts and emotions, and allows us to put a part of ourselves onto the page. This personal engagement fosters self-development in ways that artificial intelligence simply cannot replicate. Writing by hand, therefore, not only enhances learning but also nurtures personal growth, making it a practice that remains valuable, even in an increasingly digital world.

The Power Is in You

Humans don't need artificial intelligence to survive. By nature, we are intelligent beings, equipped with minds capable of extraordinary thought, boundless creativity, and incredible resilience. Just consider this: our brains are the most complex structures in the known universe, filled with mysteries that science has only begun to unravel. For all the advancements in technology, we are still far from understanding the full scope of our own mental powers. And here's the irony—in our quest to create artificial intelligence, we might be overlooking the immense, untapped potential of the intelligence we already possess. Shouldn't we be focusing on just getting to know intelligence better and developing it first, rather than artificial technology?

Your brain alone contains around 86 billion neurons, each one linking to thousands of others, forming intricate networks that let you perceive, process, and interpret everything in the world around you. We, humans, have a remarkable ability to learn, to adapt, and to reshape our thinking in response to new experiences. This adaptability—known as neuroplasticity—enables us to heal from trauma, to develop entirely new skills, and to create art and music that express our most personal emotions and ideas. Machines can replicate patterns, yes, but they miss something crucial: the emotional resonance, the subjective depth, that makes human intelligence so rich, profound, and deeply meaningful.

AI is not good primarily in three aspects: creativity, compassion, and feeling. These are the qualities that define the essence of human experience, and while AI excels in many fields, it is in these deeply personal, deeply human realms that it falls short. But this limitation, rather than being a flaw, holds the key to a better future.

Imagine a better future where people can focus primarily on things—more human. A future where the tasks and responsibilities that require emotion, creativity or sensitivity are no longer bogged down by automation and efficiency. Instead, they are elevated and cherished. AI can take care of the routine, the repetitive and the logistical—freeing up space for humans to do what only humans can do best.

Those things that give joy, that provide deep emotional connection, that require the imaginative spark of creativity, or the nuanced understanding of compassion, are the areas where humans truly shine. These qualities cannot be replicated by artificial intelligence. It cannot create art in the same way a person can, nor can it feel empathy or understand the subtle currents of human emotion in the same way we do. It cannot imagine new possibilities, new ideas, or new worlds the way a human mind can.

That's why the workplace, but also the world in general, can become better with artificial intelligence. With AI taking on the tasks it excels at, humans would be free to focus on what makes life rich: creating, connecting, feeling. The future would not be one of competition between man and machine, but one where AI supports us in doing more of what is truly human, making the world a more compassionate, creative, and emotionally fulfilling place.

To fully harness the benefits of artificial intelligence while minimizing its risks, we must remain both vigilant and proactive. It's essential to prioritize the development of strong security measures to defend against cyber threats while also investing in the preservation of our core human abilities. By fostering education that emphasizes critical thinking and problem-solving alongside technological advancements, we can build a workforce that excels at using AI yet remains capable of independent and innovative thought. This balanced approach will enable us to take full advantage of AI's potential while safeguarding our creativity and resilience for the challenges ahead.

That's why, if we're truly passionate and skilled at something, we should pursue it, regardless of how advanced AI becomes. Whether you're a lawyer, a teacher, a cashier, or in any other profession, doing what you love is the key to finding fulfillment and purpose. As AI continues to evolve, there will certainly be industries that feel the pressure of automation. However, professions that are most vulnerable to AI will still require the uniquely human qualities that machines cannot replicate—our empathy, creativity, intuition, judgment, and personal connections. These are the qualities that drive meaningful, impactful work.

While technology can assist and enhance what we do, it cannot replace the depth of human understanding and the emotional intelligence we bring to our jobs. People seek genuine human interaction, whether in legal advice, education, or customer service. No machine can replace the trust and rapport that builds in a personal relationship or the ethical decisions that professionals must navigate.

The future will need us to lean into these human values more than ever. So, instead of fearing AI, we should embrace it as a tool to enhance our work while remaining rooted in the principles that make us truly irreplaceable. In doing so, we can carve out a future where technology complements, rather than competes with, our most valued skills and passions.

About the Author

Grzegorz P. Mika is an accomplished AI leader and visionary scientist with over a decade of experience in data engineering and machine learning, driven by a passion for exploring both human and artificial cognition. As an entrepreneur, he is dedicated to developing groundbreaking solutions that push the boundaries of possibility. His extensive R&D experience in leading technology companies like ING, Samsung, T-Mobile, and Roche includes spearheading pioneering interdisciplinary projects that integrate key aspects from fields such as AI, cognitive and life sciences, and sociology, incorporating large language models into innovative commercial solutions. Currently, his research focuses on developing human-understandable explanations within social applications at the Institut Polytechnique de Paris, leveraging the potential of large language models to bridge the gap between technology and society. Additionally, Grzegorz holds a professional life coach certificate and is deeply involved with the AI Transparency Institute in Switzerland, addressing key challenges like climate change, digital ethics, AI safety, and sustainable AI.

Email: contact@grzegorzmika.com
Website: www.grzegorzmika.com

CHAPTER 19

EMBRACING GENERATIVE AI THROUGH EFFECTIVE DATA GOVERNANCE

By Salah Aldin Mokhayesh
Data Architect, Generative AI Tools
White Lake, Michigan

Ambition must be tempered by responsibility; let our generative AI respect the sanctity of data and the diversity of thought. Stand firm in guiding every model with principles of fairness and transparency. Harness the power of AI, but let it serve as an extension of human judgment, never a substitute. Establish a foundation of ethics, for it is trust that sustains progress. Reflect always that responsible governance is not a task, but a testament to the future we seek to build.
—Quote Author

The Robots Are Coming, The Robots Are Coming. But They Are Here to Help

The phrase, "The robots are coming," has long evoked both wonder and anxiety. It conjures images of a future where machines replace humans, and artificial intelligence reigns supremely. Today, it is more precise to state that robots are already integrated into our lives. Rather than signaling a takeover, they are here to support us, enhancing our work and daily activities to be more effective, intelligent, and efficient. Generative AI (gen AI), the latest evolution of artificial intelligence, stands at the frontier of this transformation. The key to utilizing this remarkable technology is understanding its role: not as a replacement for human ingenuity, but as a partner, an enabler, and an extension of human capabilities.

To grasp this concept fully, think of generative AI in the same way we think of early childhood education. A strong educational foundation is built upon a curriculum tailored to each child's needs. This curriculum is a framework—it instructs, encourages faster learning, promotes problem-solving, and ensures comprehension. The success of the learning process relies on how well this curriculum is designed and adapted to meet individual requirements. The child's ability to learn, understand, and grow is directly linked to the quality of the guidance provided.

Consider this analogy in the context of generative AI: imagine replacing the curriculum with a carefully crafted prompt, and the child with an AI model powered by a large language model (LLM) or a comprehensive knowledge base. Just as a curriculum shapes a child's learning outcomes, the quality and precision of an AI prompt determine the accuracy and relevance of its response. In this analogy, the prompt serves as the curriculum, guiding the AI to produce meaningful, insightful, and relevant outputs. Well-established documentation is vital for creating thoughtful, well-constructed prompts that effectively shape AI behavior. Thus, precise and well-documented prompts are key to achieving productive results that align with our goals and ethical standards.

The Quality of Documentation: Establishing a Solid Foundation

When considering our interactions with generative AI, the quality of the underlying documentation is paramount. Generative AI relies on vast amounts of information—policies, guidelines, standards, internal knowledge, and external resources—that must be accessible and easily understood. Documentation serves as the knowledge base for generative AI, and, much like a well-crafted curriculum, it must be meticulously designed. If information is fragmented, outdated, or poorly structured, the AI's output will inevitably reflect this lack of clarity. Therefore, we must ask ourselves: is our documentation easy to consume? Does it effectively facilitate learning for both humans and machines?

Creating high-quality documentation that is understandable, consistently updated, and accessible to both people and AI is no small task. It requires ongoing maintenance and thoughtful design, incorporating interconnected resources, regular updates, and seamless access to relevant references. Imagine a curriculum for a young student that evolves continuously, with outdated lessons promptly replaced by accurate, up-to-date content. In the realm of generative AI, documentation serves a similar purpose by providing a robust foundation to ensure the learning process—and the generation of responses—remains accurate, relevant, and insightful.

Enhancing this process involves focusing on key elements such as linking internal and external data sources, balancing static and dynamic information, and embedding documentation into systems where it is readily accessible for AI parsing and reference. Ultimately, how we create, maintain, and leverage our knowledge base determines our effectiveness in governing AI use. Quality documentation not only enables AI to deliver more precise responses but also supports responsible data governance, safeguards sensitive information, and ensures compliance with established standards.

Documentation as Governance: A Dynamic Foundation

Documentation plays a central role in responsible data governance. Its accuracy, clarity, and accessibility directly impact the effectiveness of generative AI. Beyond providing information, documentation also acts as a guiding framework, much like a curriculum shapes a student's education. Clear, well-maintained documentation allows AI to understand queries better and produce responses that are both accurate and relevant.

The importance of dynamic documentation cannot be overstated, it is essential that documentation evolves alongside changing organizational needs and new information. Static information can quickly become obsolete, leading to inaccuracies, confusion, and misalignment. Conversely, dynamic documentation that is updated regularly and enriched with internal and external links serves as a living resource, reflecting the latest policies, guidelines, and best practices. This ensures that AI can reference up-to-date information, delivering responses that are consistent with current standards.

For instance, consider an organization with internal guidelines for managing sensitive customer information. These guidelines are embedded in a central knowledge base and updated in real time whenever regulatory requirements change. By referencing these updated guidelines, generative AI ensures that responses and actions involving customer data remain in full compliance. This seamless integration between dynamic documentation and AI actions upholds governance standards while minimizing the need for constant manual oversight.

Data Governance: Defining Our Principles

Data governance is more than just a set of rules and guidelines; it is the compass that directs the ethical and responsible use of generative AI. The way we govern data shapes our reputation, dictates how we address biases, and defines how we maintain accuracy, privacy, and public trust. The critical question is not whether we should use generative AI—its presence is already established and here to stay. Instead, the focus is on how we can use it responsibly, ensuring our actions align with the best interests of individuals, society, and the ethical standards we uphold.

The first step in establishing effective data governance for generative AI is setting clear goals and expectations. This can be accomplished by forming a governing body and council akin to a data governance council—that is dedicated to safeguarding ethics, addressing biases, and ensuring privacy and safety. An "AI council," as it could be called, would be tasked with setting these expectations and continuously monitoring, adapting, and evolving the guidelines as the technology progresses.

Generative AI and data governance must operate in tandem. Generative AI is a powerful tool designed to enhance, accelerate, and improve productivity, but it should not be relied upon to independently generate ideas, exercise creative thought, or guarantee impartiality. Rather, AI should function as an extension of our capabilities as a tool that amplifies human ingenuity while keeping the decision-making firmly in our hands. Simply put, AI is here to assist, not to replace.

AI and Tedious Tasks: Reducing Cognitive Load

Generative AI is particularly effective for managing tedious, repetitive, and time-consuming tasks, allowing human workers to focus on areas that require creativity, empathy, and strategic thinking. By utilizing generative AI, organizations can enhance the quality and efficiency of processes such as creating and managing quality rules, testing those rules, assisting with engineering tasks, supporting data stewardship, enriching metadata, and establishing and enforcing compliance standards.

Imagine having a personal assistant that excels at reviewing large volumes of documents, extracting key insights, or summarizing lengthy reports. This is precisely the role that generative AI fulfills. It functions as a highly capable assistant, able to scale rapidly to meet organizational demands without experiencing fatigue or overload.

When AI operates within a well-maintained knowledge base, it can draw on a wealth of insights, best practices, and contextual knowledge that reflect the collective intelligence of an organization. This enables generative AI to provide accurate, consistent, and well-informed responses that align with organizational standards. Take data profiling, for example: AI can efficiently analyze extensive datasets, identify patterns, and flag anomalies. It can classify data based on its sensitivity, ensuring compliance with regulatory standards. By referencing a knowledge base that clearly defines sensitive data, generative AI can automatically apply data protection rules, preserving privacy and minimizing the risk of exposure.

Improving Data Quality Through AI

Generative AI offers substantial improvements in data quality management, serving as a key component of modern data governance. One of its notable capabilities is the generation of synthetic data, which plays a crucial role in training models and testing applications while maintaining user privacy. This synthetic data is particularly valuable for evaluating model performance, refining metrics, and enhancing the effectiveness of data science models. By using synthetic data, organizations can simulate real-world scenarios, assess model robustness, and optimize processes without compromising sensitive information. This approach not only addresses privacy concerns but also contributes to more secure and effective model development.

In addition to enhancing data quality, generative AI provides a consistent and sophisticated solution for data cleansing, which is especially beneficial for managing large, complex datasets. As a fundamental aspect of comprehensive data governance, effective data cleansing involves identifying and rectifying errors, resolving inconsistencies, and ensuring that all retained data is accurate, relevant, and high-quality. As data volumes increase, manual data cleansing becomes increasingly inefficient and error-prone, leading to incomplete, inaccurate, or misaligned data.

Integrating generative AI into data governance frameworks allows organizations to automate the cleansing process, ensuring consistency, accuracy, and reliability across all datasets. This technology can identify anomalies, detect discrepancies, and adapt to evolving data patterns in real time, making it indispensable for ongoing data monitoring and maintenance. Moreover, generative AI can analyze data flows from source applications, pinpointing issues such as poor data quality or incorrect formats. Once identified, the AI can propose or implement corrective measures, such as standardizing formats, identifying duplicates, or recommending adjustments to data collection methods. This proactive and automated approach not only strengthens analytics and insights but also supports more informed decision-making by providing a solid, reliable foundation of high-quality data.

The application of generative AI in data governance illustrates the importance of ethical usage and robust oversight. While AI technology is increasingly integrated into organizational processes, its impact is inherently shaped by how it is designed, implemented, and governed. Responsible AI governance is crucial to transforming generative AI from a potential risk into a powerful asset that drives operational efficiency, enhances data accuracy, and fuels innovation. Through well-implemented frameworks, generative AI can elevate data-driven initiatives, ensuring both compliance and long-term strategic value.

Establishing Trust: Ethical and Responsible AI Use

Trust is the cornerstone of effective AI use, and trust must be built upon ethical and responsible data governance. Organizations that fail to govern their AI systems responsibly risk losing the trust of their customers, stakeholders, and the public. The path to trustworthy AI begins with a commitment to transparency, fairness, and ethical practices along with principles that must be reflected in both AI implementation and data governance.

An AI council, dedicated to overseeing the ethical use of generative AI, plays a crucial role in establishing and maintaining this trust. The council's responsibilities would include identifying potential biases, evaluating AI use cases for fairness, and implementing protocols to protect privacy. Bias is one of the most significant challenges facing AI today. The datasets used to train AI models often reflect existing biases, which, if not addressed, can lead to biased outcomes. An AI council can guide efforts to mitigate these biases, ensuring that AI models are trained on diverse datasets that accurately represent the world in all its complexity.

Moreover, the ethical use of generative AI requires adherence to safety protocols and guidelines that prevent AI from being used in ways that could harm individuals, communities, or society. Generative AI should be viewed as a tool that supports human effort, not as a decision maker. By adhering to safety protocols and establishing clear boundaries for AI use, organizations can harness the benefits of generative AI while safeguarding against its potential risks.

Governing AI Responsibly for a Better Future

The robots are not coming—they are already here. The critical question is not whether we can stop them (we cannot), but how we can govern their presence in a way that aligns with our values, priorities, and ethical standards. Generative AI holds immense potential to enhance productivity, improve decision-making, and create value, but this potential can only be realized if we take responsibility for how it is used.

As we embrace AI, we must stay focused on what defines us as human: our capacity for ethical reasoning, empathy, and creativity. When governed responsibly, generative AI has the power to amplify these qualities, creating a future in which technology serves humanity rather than controls it. It is up to us to lay the foundation that enables AI to help build a better, fairer, and more innovative world.

About the Author

Salah Aldin Mokhayesh is a data architect with a focus on data governance, data architecture, and generative AI. He holds a mathematics degree from Eastern Michigan University and has developed solutions across his field. Salah is an experienced mentor and an aspiring author, with a passion for sharing insights on data strategy and technology. He also holds certifications in leadership, generative AI, data governance, and architecture.

Email: salahmokhayesh@gmail.com

CHAPTER 20

BALANCING INNOVATION AND ETHICS: NAVIGATING THE DATA-DRIVEN AI REVOLUTION

By Louis Mono
Lead AI Engineer, PhD candidate in AI, Musician
Milan, Italy

The earth has enough resources to meet everyone's needs, but not enough to satisfy everyone's greed.
—Gandhi

Does AI Really Exist?

The machine is programmed to give us the illusion that it is not a machine. If we adhere to this statement, it becomes apparent that humans, since the dawn of time, have been designing and programming tools to automate tasks they already know how to perform. Within this framework, we can see that artificial systems have been created by humans for centuries. What distinguishes the current era—beginning with Turing's tests in the 1950s, the creation of the internet, the official launch of the

World Wide Web in the early 1990s, and the explosion of various social media platforms in the 2000s—is unprecedented access to vast amounts of human data. From consumer behavior and search histories to health records, political opinions, and sensitive personal information, the cataloguing of personal data for mass surveillance has become a widespread practice. In reality, nothing you say on the phone, no website you visit, no email or message you send or receive, and no post you share on social media is truly private. These pieces of information are often stored by tech companies operating across various sectors, and their data banks are accessible to any entity or individual with authorized access.

Data is often described as the oil of the 21st century, highlighting its immense value and the critical role it plays in today's digital economy. Therefore, artificial intelligence (AI) today can be defined as the development of systems capable of performing tasks that traditionally require human intelligence, by learning from vast amounts of data. These systems rely on past experiences and data to continuously enhance their performance, enabling them to replicate human decision-making and behaviors.

AI is generally categorized into three main types: narrow, general, and super. Narrow AI, also known as weak AI is designed to excel in specific tasks such as facial recognition or internet searches. Even models like OpenAI's ChatGPT fall under this category, as they are limited to a singular function, such as text-based conversation. General AI or strong AI aspires to surpass human intelligence across a wide range of cognitive tasks. Unlike narrow AI, general AI can apply its acquired knowledge and skills to new situations without needing human retraining or intervention. Super AI, a theoretical concept, is often referred to as artificial superintelligence. If realized, super AI would possess reasoning, learning, and cognitive abilities that far exceed human capabilities, allowing it to think, make complex judgments, and innovate in ways beyond human comprehension.

But can we truly understand an individual through a collection of data? From a spiritual and holistic perspective, we can recall Albert Einstein's assertion, "Everything in the universe is vibrational." Human beings are far more than their behaviors, actions, habits, or the billions of data points collected about them. In this light, despite society's efforts to reduce us to mere information, as philosopher Merleau-Ponty observed, "Another's behavior is not the person." True artificial intelligence, then, cannot exist without understanding our vibrational intelligence; it would merely be an extension of that understanding.

When humans come to realize that they are, above all, energy—and learn to control their frequencies—we could see polymorphic flying machines commanded by thought, healing through the living energy emanating from our hands, and instantaneous travel through mere intention. A new, more conscious species would emerge, one that could repair, build, and reshape matter throughout the biosphere, guided by the intention of the heart, thoughts, and vibrations.

Given all of this, can we truly say that artificial intelligence already exists? According to the current materialistic paradigm, can the artificial systems we develop through programming—mere amplifiers of human expertise—genuinely be considered "intelligent"? Can we even speak of intelligence if these systems lack self-awareness?

Beyond the practical benefits and solutions that artificial systems provide to society, what moral framework governs their responsible and ethical use? Will we allow AI to evolve in the future without ethical constraints, potentially engaging in conflict to the point of threatening our planet's survival—much like the state of the world today, teetering on the brink of a third world war with nations neutralizing each other under the shadow of nuclear threats?

Practical Uses of AI in Our Societies

In a constantly evolving technological landscape, the interdependence between data and artificial intelligence has emerged as a critical driver for growth and innovation within businesses. Organizational leaders are at the forefront of this revolution, leveraging the potential of data and AI to propel their companies toward success. For example, in the medical field, AI systems enhance efficiency and productivity in cancer diagnoses. Research by Scott McKinney and his team on breast cancer diagnosis demonstrated that using an AI system to interpret mammograms resulted in a 5.7% reduction in false positives and a 9.4% reduction in false negatives (2020). These AI tools improve accuracy, reduce costs, and save time compared to traditional diagnostic methods.

Moreover, AI has the capacity to process vast amounts of data at remarkable speeds, enabling real-time analysis and insight generation. This significantly accelerates and streamlines decision-making processes, especially when automation is integrated across various stages of the workflow. A prime example is recommender systems and social computing, which, within marketing, allow professionals to better understand the social behaviors and dynamics of a target market, thereby improving decision-making.

The impact of AI extends beyond immediate applications and into futuristic technologies. AI acts as a catalyst for innovation, driving advancements in fields like transportation, particularly with self-driving cars. These vehicles have the potential to improve road safety, reduce traffic congestion, and increase accessibility for individuals with disabilities or mobility challenges. Additionally, the use of AI-powered robots is revolutionizing industries such as manufacturing—where robots assist with stacking, assembling, and even daily household tasks. In high-risk situations, AI robots offer distinct advantages over humans. They can be deployed in dangerous environments, such as space exploration, deep-sea expeditions, coal and oil mining, or even mitigating the effects of a bomb by controlling fires in their early stages. Whether navigating natural disasters or man-made crises, AI can address hazards that would otherwise limit human intervention.

Another key advantage of AI is its round-the-clock availability. AI-driven digital assistants and customer support chatbots offer instant assistance to users anytime, anywhere. This enhances user experience and provides continuous service without the limitations of human schedules.

Finally, in the realm of augmented reality, AI is paving the way for practical innovations like real-time translation through smart glasses. For instance, with Meta glasses, users will be able to converse with someone speaking French or Italian and hear the translation in their chosen language via the glasses' open-ear speakers. Despite the philosophical concerns surrounding the metaverse, such a feature could be a practical solution for fostering communication between people of different linguistic and cultural backgrounds.

In sum, AI holds immense potential across multiple sectors, from healthcare to everyday tasks, transforming how we interact with technology and shaping the future of society. However, this evolution must be accompanied by thoughtful consideration of its ethical implications and societal impact.

Ethical Considerations in AI Development

The symbiotic relationship between data processing and AI-driven automation, which lies at the heart of technological innovation, seems to prioritize profit over human-centered progress. But what if ethical AI development simply meant placing human beings at the center? What do we mean by "ethics" in this context? Who defines these ethical standards, and what is their ultimate purpose? Do they respect the diversity of human existence, the natural balance of ecosystems? Or are they a selective, hierarchical construct, where dominant forces—multinational corporations—subjugate smaller entities, perpetuating a cycle of masters and slaves? When examining issues like mass surveillance, lack of transparency, and human-machine integration, it becomes clear that many principles of human dignity are being compromised.

In the context of privacy and surveillance, for example, the widespread use of AI-driven technologies raises profound concerns. While facial recognition technology offers security benefits, it also risks being misused for mass surveillance, infringing on personal privacy rights. The threat of data breaches from AI-powered surveillance systems also endangers personal and financial information, leading to identity theft and other forms of cybercrime. In the defense industry, ethical challenges are even more pronounced. The rise of robotic dogs and war drones presents the troubling scenario of machines being programmed to take human lives. Delegating the decision to kill to an autonomous system erodes the very essence of human dignity. Though machines may inadvertently cause accidental deaths, intentionally programming them to kill strips away the moral value of human life, reducing it to a mere calculation.

Furthermore, bias and discrimination are other critical concerns. AI systems often reflect the biases inherent in their training data, sometimes exacerbating these biases. This can lead to unjust outcomes, such as favoring certain demographic groups in hiring processes or perpetuating discrimination in various fields. Additionally, the transparency of AI systems is frequently compromised. Complex deep learning models often function as opaque "black boxes." This undermines accountability and trust when errors or harmful outcomes occur. This lack of transparency raises questions about the moral responsibility of advanced AI systems, especially in cases like hallucinations in ChatGPT. In the context of creativity, AI may be capable of generating art, music, or writing, but it lacks genuine originality or the ability to think outside the box. AI's reliance on existing data patterns constrains its ability to produce creative work that is truly innovative, as it cannot replicate the intuition or emotional intelligence that fuels human creativity.

The transhumanist ideology, particularly the concept of the singularity predicted by American computer scientist and AI expert Ray Kurzweil to occur post-2045, suggests that humans may eventually merge with machines to avoid extinction as artificial intelligence surpasses human cognition. This vision foresees Humans 2.0—patented, licensed, and controlled beings—capable of uploading their consciousness to computers. Through advancements in genetics, nanotechnology, and robotics, human biology would be governed by technology. Natural reproduction would be replaced by artificial wombs, and sex as we know it would cease to exist.

But where does this leave human dignity? What appears to be progress for the masses could be nothing more than a sophisticated form of imprisonment, the end of free will. Is this the future that artificial intelligence is leading us toward—controlling our minds, dictating what to think, how to think, and when to think, all in the pursuit of fragmented instantaneity? By fostering a fear of death and offering the illusion of material eternity, is AI actually enslaving humanity? In this paradigm, at what point does a human-machine hybrid break free from the matrix to reclaim true freedom? How can we, as humans, hope to merge with machines or claim to understand AI more deeply than we understand ourselves, especially when we have yet to fully grasp our own vibrational intelligence, energy, and frequencies?

Given these ethical complexities, it is essential to establish a responsible framework for AI development. International laws must be designed to govern our relationship with advanced systems in a way that avoids perpetuating cultural biases, maintains transparency, and resists extreme measures that seek to control human life. Technology should serve humanity, not diminish it; we must not sacrifice free will or force integration with machines at the expense of our core human values.

Taming AI: Bridging the Spiritual and Material for a Harmonious Future

Between those who wholeheartedly embrace technology without questioning the serious risks associated with artificial intelligence—such as the erosion of free will, the emergence of autonomous weaponry, the proliferation of manipulative public opinion, the spread of misinformation, and the creation of deceptive imagery—and those who categorically reject AI advocating solely for an ecological approach, I want to argue that our evolution lies in both directions. We must bridge the spiritual and material realms, adopting a holistic and scientific approach simultaneously. The objective is not to sanctify AI or to oppose it outright; rather, it is to "tame AI," much like the Little Prince tamed his rose in Antoine de Saint-Exupéry's narrative.

To effectively navigate the inevitable upheaval brought about by AI, I advocate for one fundamental principle: everything is energy—present in nature and inherent in every human being. The intricacies of life extend far beyond any singular ideological viewpoint. Undoubtedly, society will experience significant transformations; however, the fundamental law of energy remains unchanged: nothing is gained, nothing is lost; everything merely transforms. By observing nature through the lens of biomimicry and respecting all elements that contribute to ecological balance, we can aspire to usher in an era of super AI through a genuinely holistic and scientific approach.

This synthesis could lead to the emergence of a new human species that harmoniously combines vibrational intelligence with a healthy form of artificial intelligence. By integrating biomimicry with advanced automation, we have the potential to create a new civilization founded on principles of sharing and harmony.

We Must All Address This Question: Can AI Truly Be Ethical?

The question of whether AI can truly be ethical is a complex one, and providing a definitive answer is far from simple. Morality, after all, is a multifaceted concept, shaped by our diverse socio-cultural contexts and the intricate identities we carry—even within a single nation. Different regions of the world develop their own ethical frameworks, each inherently subjective to its own perceptions of right and wrong. Yet, the more urgent question is whether there can ever be a system that integrates all these varied ethical structures. The current Western system, in particular, is deeply lacking in this regard and stands on the brink of crisis. Over time, the Western world has shifted toward an increasingly individualistic, capitalist, and materialistic model. This has created a life-draining matrix that constrains human potential and, more crucially, starves our spiritual well-being.

As René Guénon argued in *The Crisis of the Modern World,* the 21st century must be spiritual, or it risks not existing at all. Guénon believed that modernity represents a period where the essence of being is forgotten, replaced by an unrelenting focus on action and materiality. This focus turns people into mere components of the material world. He described the modern era as one characterized by a constant need for activity, change, and speed—reflecting the rapid progression of events around us. This era of modernity has led to a fragmentation of life into multiplicity. The unity of purpose and higher principles that once guided humanity has been lost, replaced by an obsession with analysis and division. Human activity, in this state, is disjointed, dispersed across numerous domains, resulting in an inability to synthesize and focus—a trait that Guénon suggested strikes Eastern thinkers as particularly troubling. These patterns of behavior are the inevitable consequences of an ever-deepening materialism.

Materiality, by its very nature, fosters division and conflict, both on a global scale and within the human psyche. The more we sink into material concerns, the more the forces of opposition and division intensify. Conversely, Guénon argued, ascending toward pure spirituality brings us closer to unity—an ideal only achievable through an awareness of universal principles.

Considering this, it seems highly probable that remaining within the current system will make it incredibly difficult to control a super AI, should one ever be developed. How can we genuinely speak of ethical AI if, in the end, it is governed by the laws of money and profit, driven by the most negative aspects of these forces? To navigate this challenge, I believe the entire system must be reimagined. We must move toward a new paradigm that emphasizes values such as solidarity, sharing, and compassion. A more equitable order is required across political, economic, and social realms—one grounded in spiritual consciousness and universal truths, respecting the natural balance and restoring human dignity.

No matter what the future holds, it is essential to remember that the evolution of AI is within our hands, and its trajectory is not deterministic. We must turn inward, cultivate self-awareness, and guide AI development with heartfelt intention. The greatest revolution lies in our consciousness—the profound journey of truly knowing ourselves. And with that, I leave you a fragment of my soul, which, at its core, is inherently poetic.

No masters, no slaves
All living beings, without enclaves
Called by the voice of the soul
Lit by the echo of flames

From drops to eternity
Time does not exist. We are bound by Love.
Here to Be. To create our infinite realities.

About the Author

Louis Mono is a poetic soul, a multifaceted individual whose talents span both science and art. At 30, he published his debut work, *La Première Question, Tome 1* in 2020, which became a bestseller in the Francophone world. In 2021, he released an expanded edition of the same book, followed by a new publication in 2023 titled *De l'intelligence artificielle à l'intelligence vibratoire*. In addition to his literary work, Louis is a musician, having produced the album *Chlodésie, Partie 1: VITЯIOL*, along with several singles available on digital platforms.

Louis is an engineer specializing in artificial intelligence. His expertise goes beyond theoretical discussions or philosophical views on AI. He has a deep understanding of the algorithmic and technical frameworks that drive artificial systems. Now pursuing a PhD in AI, he is exploring ethical issues and examining the relationship between humans and these emerging technologies. He advocates for an approach that recognizes human evolution as a blend of spiritual (vertical) and material (horizontal) development. A seasoned consultant, Louis also engages in coaching and teaching, guiding others with his expertise.

Looking ahead, Louis is working on various projects in art and AI, aiming to foster a new paradigm that bridges the spiritual and material worlds. As he reflects, "The cause of every creation is lost in the infinite; the line stretches from so far away—if only our veiled eyes could finally see."

Email: louis.mono1@gmail.com
Website: https://linktr.ee/louismono

CHAPTER 21

INNOVATING FOR A SUSTAINABLE FUTURE: AI AND DATA-DRIVEN SOLUTIONS

By Greg Ombach, PhD
Senior Executive, Driving Transformation, Innovation, Growth
Munich, Germany

Intelligence is whatever machines haven't done yet.
—Larry Tesler, 1970

Introduction: Transforming Industries with the Power of AI

Throughout my career at the forefront of innovation and sustainability, I have witnessed how AI and data-driven strategies are not just tools but transformative forces, unlocking groundbreaking solutions to some of humanity's most urgent challenges. From creating more intuitive user experiences in consumer electronics to enabling autonomous vehicles and optimizing aerospace operations, AI is becoming a cornerstone of industry transformation.

Human and machine intelligence are fundamentally different yet deeply complementary. While humans excel at creativity, judgment, and empathy, machines bring unmatched capabilities in processing vast datasets, identifying patterns, and executing tasks precisely. These strengths enable innovative solutions that neither humans nor machines can achieve alone.

Generative AI is exciting with its ability to analyze data, identify patterns, and generate new designs and models based on existing knowledge. Yet, as recent statistics suggest, its adoption is still in its early stages. Many organizations remain in pilot phases, cautious about scaling experiments despite the transformative potential. Moving beyond the "hype cycle" toward tangible productivity requires practical and scalable applications that deliver measurable outcomes.

As history has shown with past technological revolutions, today's cutting-edge AI will evolve into foundational systems—smart operations, smart applications, or workflows embedded seamlessly into everyday tools. Over time, what we now call AI will transform into standard, automated capabilities that integrate into our lives without notice.

While embracing AI, we must also navigate challenges like data privacy concerns, energy consumption, and ethical questions like explainability and hallucinations. Generative AI models consume significantly more power than traditional technologies, raising sustainability concerns as adoption scales. Addressing these requires not just technical innovation but also frameworks for responsible implementation.

This chapter will explore how AI and generative AI are revolutionizing three key sectors: consumer electronics, automotive, and aerospace. Through real-world examples and use cases, I will illustrate how AI-driven innovation creates more intelligent systems, drives sustainability, and enables breakthroughs that align with the broader goals of human progress. Let us uncover how AI reshapes industries and lays the foundation for a more sustainable and inclusive future.

Consumer Electronics: Revolutionizing Everyday Experiences

Artificial intelligence (AI), particularly generative AI, is transforming the consumer electronics sector by making devices more intuitive, efficient, and personalized. This section explores how these innovations reshape how we interact with technology and the implications for everyday life.

An important innovation is the move toward on-device AI, which involves deploying smaller, task-specific models directly on devices. Unlike larger models such as LLMs, these focused systems are tailored for efficiency, making them faster, less resource-intensive, and better suited for real-time applications. By reducing dependence on remote data centers, on-device AI enhances performance, lowers energy consumption, and supports sustainability goals. This approach allows AI algorithms to function locally on devices, enabling quicker responses while minimizing the environmental impact of extensive cloud computing.

Generative AI also redefines the user interface, enabling people to interact with devices using natural language. Users can ask their devices to perform tasks instead of navigating complex menus or apps. For instance, imagine saying, "Can you book dinner with my friend for Saturday night at our favorite Italian place?" The device would manage everything, from checking schedules to making reservations, seamlessly integrating into daily life.

AI agents are central to these advancements—intelligent systems designed to understand user needs, gather relevant information, and make decisions to complete tasks efficiently. Unlike traditional apps that operate independently, AI agents act as intermediaries, connecting various tools and services to deliver cohesive solutions.

This marks a shift in user interaction from traditional app-based models to AI-first experiences. Instead of relying on siloed applications, interactions become fluid and integrated. AI agents are now moving from operating in vertical silos—limited to individual apps or functions—to working more horizontally, connecting multiple apps and systems seamlessly. This ability to integrate diverse applications enhances the user experience by streamlining tasks and making interactions more intuitive.

Generative AI also advances spatial computing technologies like smart glasses and augmented reality. These devices leverage smaller, specialised AI models to process visual and auditory inputs in real time, enabling tasks like translating signs, providing directions, or offering contextual information about the user's surroundings.

Another critical development is the ability of devices to integrate multiple specialized AI models simultaneously. For example, a device might use one model for natural language processing, another for visual recognition, and another for understanding user preferences. This combination creates more context-aware interactions tailored to the user's specific needs.

Whether booking dinner with a friend or navigating a foreign city with augmented reality glasses, Generative AI is making technology more accessible, efficient, and deeply integrated into our lives. AI agents, capable of seamlessly connecting and managing various apps and systems, are at the forefront of this transformation.

Automotive Sector: Redefining Mobility Through Intelligent Systems

Building on the advances in consumer electronics, generative AI is reshaping the automotive sector with innovations that enhance usability, safety, and efficiency. Just as AI agents connect and manage apps seamlessly in personal devices, they are revolutionizing in-car systems to provide smarter, safer, and more personalized interactions. By leveraging consumer electronics processing and smaller on-device AI models, the automotive industry is driving a new era of intelligent mobility that touches every aspect of the value chain, promising an exciting future of mobility.

Generative AI enhances the customer experience by personalizing every aspect of the in-car environment. Conversational assistants and voice controls now allow drivers to perform tasks hands-free, improving convenience and safety. Adaptive settings for climate, seat positions, and music preferences ensure that each journey feels tailored to the individual's preferences. Digital cockpits take personalization further by integrating advanced technologies like augmented reality displays, gesture recognition, and emotion prediction. These systems transform vehicles into highly responsive environments, offering real-time insights, personalized navigation, and enhanced driver assistance features, making the overall driving experience intuitive, engaging, and safer.

Autonomous driving is one of the most important uses of generative AI in the car industry. Developers use AI to create synthetic data and simulate different driving situations, testing self-driving systems in safe virtual environments. This helps vehicles handle rare or extreme conditions, like sudden weather changes or unexpected obstacles, which are hard or dangerous to test in real life. Generative AI improves how self-driving cars make decisions, making them more reliable and bringing us closer to using them widely.

A significant advancement in AI is its transition from operating purely in digital realms to interacting with and manipulating the physical world. This shift, often called "AI getting physical," enables AI-driven humanoid robots to perform tasks in real-world environments, from homes to industrial settings. In the automotive sector, these robots are revolutionizing production lines, managing tasks such as assembling components and handling heavy materials. By integrating reinforcement learning and edge AI, these robots adapt to new workflows and navigate human-centric environments seamlessly, making them indispensable tools for enhancing productivity and human-robot collaboration.

Whether personalizing the driving experience, optimizing vehicle design, or enhancing production efficiency, generative AI is proving to be a transformative force in the automotive sector.

Aerospace Sector: Elevating Innovation for a Sustainable Future

Like the automotive industry, aerospace is undergoing a profound transformation driven by sustainability goals and digitization. Decarbonization by 2050 marks the fourth revolution in aviation, following the achievements of flight, safety, and accessibility. Today, the focus is on making aviation sustainable, with generative AI playing a critical role in reshaping how the industry approaches design, manufacturing, operations, and supply chain management.

Generative AI is making significant contributions across every stage of the aerospace lifecycle. During the design phase, AI accelerates innovation by analyzing extensive datasets and generating optimized aerodynamic structures, reducing drag and improving fuel efficiency. This enables the development of lighter and more sustainable aircraft. AI-powered generative design tools create multiple component prototypes in record time, allowing engineers to identify the best designs that meet rigorous performance and safety criteria. This shortens the design cycle and supports the creation of fuel-efficient aircraft tailored for future air traffic demands.

AI enhances manufacturing processes by streamlining workflows, improving precision, and reducing material waste. For example, AI-driven robotics in assembly lines ensure consistent quality while accelerating production rates. Automation supported by generative AI also creates more intelligent factories, where processes like component fitting, testing, and adjustments are carried out with minimal human intervention, reducing errors and enhancing scalability.

Operational efficiency is another crucial area where generative AI delivers value. A prime example is the Airbus A350, which collects approximately 2.5 terabytes of data daily during its operations. This data provides valuable insights for predictive maintenance systems, which analyze real-time information to identify potential issues before they become critical. These AI-powered systems improve safety, reduce downtime, and enhance operational reliability. Predictive AI also helps extend the aircraft's lifespan and reduces unplanned disruptions, e.g., "aircraft on the ground" (AOG), by ensuring timely and efficient maintenance.

Generative AI is also enhancing flight performance and sustainability. For instance, data collected from weather monitoring, fuel flow, and engine health can be analyzed to optimize flight paths, resulting in significant fuel savings and reduced emissions and contrails. Such AI-driven insights directly impact sustainability goals, ensuring that airlines meet regulatory standards and contribute meaningfully to decarbonization efforts.

Generative AI optimizes logistics in supply chain management by forecasting demand, managing inventory, and improving supplier coordination. AI-powered solutions can analyze historical and real-time data to streamline the delivery of parts and components, ensuring that production schedules remain uninterrupted. This efficiency reduces costs, minimizes waste, and lowers the carbon footprint of supply chain operations, contributing to sustainability across the value chain.

Looking to the future, generative AI is poised to address even more complex challenges. Collaboration across industries could lead to innovations such as advanced coatings that resist wear and corrosion, extending the durability of critical aircraft components. Similarly, AI-driven autonomous systems are expected to enhance navigation, optimize flight operations, and reduce pilot workload while maintaining stringent safety standards. Cross-industry applications of these advancements, such as integrating AI with adaptive logistics platforms, could create more efficient global supply chains benefiting multiple sectors beyond aerospace.

The potential of generative AI becomes even more exciting when combined with emerging technologies like quantum computing. While generative AI provides predictive and analytical capabilities, quantum computing can unlock unprecedented processing power for solving highly complex problems, such as simulating aerodynamic flows or optimizing energy usage. These combined technologies could redefine the aerospace industry's ability to innovate at scale and with unparalleled precision.

Generative AI transforms aerospace by optimizing design, automating manufacturing, enhancing operations, and streamlining supply chains. It drives safety, efficiency, and sustainability while paving the way for future innovations like autonomous systems and advanced logistics. This marks a pivotal step toward a more intelligent aviation industry integrating cutting-edge technology with responsibility and sustainability.

Conclusion: Pioneering the Era of Augmented Regenerative Intelligence

Artificial intelligence (AI) and data-driven strategies are reshaping industries globally, and this chapter highlights a few examples from my experience in consumer electronics, automotive, and aerospace. In consumer electronics, integrating on-device AI and generative AI has transformed user experiences through natural language interactions and personalized services, making technology more intuitive and accessible. In the automotive sector, generative AI enables the personalization of in-car environments, advancing autonomous driving capabilities, optimizing vehicle design and manufacturing processes, and driving improvements in efficiency and sustainability. Generative AI is revolutionizing aircraft design, manufacturing, and operations in aerospace by analyzing extensive datasets to develop more efficient and sustainable aircraft, enhancing safety and reducing environmental impact.

Generative AI will support innovation by enhancing creativity, speeding up problem-solving, and enabling breakthroughs across industries. For example, AI-powered generative design tools can analyze vast datasets to propose optimized solutions for challenges like lightweight materials or aerodynamic efficiency. These tools enable organizations to experiment rapidly, accelerating innovation cycles by creating and testing virtual prototypes in hours instead of months. Moreover, generative AI's ability to uncover hidden insights in complex data opens new pathways for innovation, finding opportunities that may have previously gone unnoticed.

However, challenges remain to be addressed before generative AI can reach broad adoption, in safety-critical and non-safety-critical systems. Issues such as regulations, standardizations, explainability, and mitigating hallucinations must be resolved to ensure these technologies' safe and ethical deployment. As industries adopt AI at an accelerating pace, the need for robust frameworks that provide reliability, transparency, and compliance is more pressing than ever. This adoption cycle is significantly faster than previous technological shifts, meaning organizations that fail to integrate AI risk falling behind in competitiveness.

Despite these challenges, generative AI's potential to enhance human capabilities and transform industries remains vast. Integrating large language models (LLMs), small language models (SLMs), and edge AI in robotics is already driving advancements in efficiency, accuracy, and overall functionality.

Realizing AI's transformative potential and overcoming its challenges will depend also on meaningful partnerships between established companies and dynamic startups, enabling innovation and accelerating the integration of new technologies. An impactful collaboration model based on a venture client unit (a corporate department that works with startups to improve a company's competitiveness) has proven highly effective in adopting startup solutions within enterprises.

Looking ahead, the new concept of augmented regenerative intelligence (ARI) provides a promising vision for the future of technology. By synthesizing natural, human, and artificial intelligence, ARI seeks to create adaptive, self-healing, sustainable systems that benefit all life forms and ecosystems. This approach goes beyond solving immediate challenges, aiming to establish a harmonious balance between technological progress and the natural world. By embracing this vision, we can ensure that innovation drives progress while fostering a more equitable and resilient future where technology supports humanity and the planet.

About the Author

Dr. Grzegorz (Greg) Ombach is a senior executive with a track record of leading deep-tech innovation and scaling businesses for global impact. With leadership experience at Siemens VDO, Qualcomm, and Airbus across Europe, the USA, and China, he has successfully transformed cutting-edge technologies into market-leading solutions.

As Senior Vice President and Head of Disruptive Research and Technology at Airbus, Greg drives the development and commercialisation of breakthrough technologies, forging strategic partnerships that redefine industries. His ability to build and lead high-performing teams, accelerate business growth, and create sustainable competitive advantages positions him as a transformational leader in the deep-tech space.

In addition to his corporate leadership, Greg serves as a chairman, board member, and advisor to startups, scale-ups, and public companies, helping them navigate digital transformation, AI, and sustainability challenges. His expertise in innovation strategy, business creation, and market adoption makes him a visionary leader ready to take deep-tech companies to new heights.

Email: greg@navinn.tech
Website: www.navinn.tech
LinkedIn: https://www.linkedin.com/in/grzegorz-ombach/

CHAPTER 22

AI IN SYSTEMS ARCHITECTURE: THE BEAUTY OF SIMPLICITY

By Ivan Padabed
Systems Engineering Methodologist
Loulé, Portugal

> *Simplicity is the ultimate sophistication.*
> —Leonardo da Vinci, the original Renaissance engineer

This chapter will explore how AI plays a transformative role in the daily lives of engineers. We will dive into three core areas where AI doesn't just assist but amplifies human potential, offering solutions to the ever-growing complexities of modern engineering. Each section highlights AI's unique capabilities as a cognitive partner, expert consultant, and reflective mirror for engineers, driving efficiency, accuracy, and balanced decision-making.

1. *Cognitive Power and Forecasting Capabilities*
In this section, we explore AI's ability to amplify human cognition. Engineers face an overwhelming flow of information in the modern world, and filtering what's essential can be a monumental challenge. AI

serves as an "exocortex," like an extension of the human brain, adding elaborated details to human-generated high-level concepts, ensuring internal consistency of the models and their overall alignment with an overarching vision.

2. *Precision, Expertise, and Responsibility*
AI is like your "personal SWAT team," instantly connecting you to domain expertise. From prioritizing usage scenarios to identifying missing details that could cause problems later, AI ensures engineers don't overlook anything important. This section highlights AI's attention to detail, domain knowledge, and its ability to act responsibly by maintaining accountability throughout project lifecycles.

3. *Reflection and Bias Elimination*
The final section discusses how AI acts as an engineer's "alter ego." It aids in "self-reflection," boosting an engineer's self-confidence and identifying opportunities to improve by offering unbiased perspectives. By eliminating cognitive biases and removing emotional influences from decisions, AI ensures that engineering choices remain logical and grounded in reality.

1. Cognitive Power and Forecasting Capabilities

> *An engineer, a physicist, and a mathematician are asked to prove that all odd numbers are prime. The physicist checks a few odd numbers and declares: "3 is prime, 5 is prime, 7 is prime, 9 is an experimental error, 11 is prime, so it must be true for all odd numbers." The mathematician writes a formal proof showing that not all odd numbers are prime, pointing to 9 as a counterexample. The engineer takes one quick look at 3, 5, 7, 11 and says, "Good enough for my purposes!"*
> —Traditional engineering humor, author unknown

Pragmatic Engineering: Why "Good Enough" Beats Perfection
Engineers are the ultimate problem solvers, but they're also realists. In the fast-paced, resource-constrained world of engineering, perfection isn't the goal—it's usually a luxury. Instead, the focus is on delivering "good enough" solutions to get the job done. This doesn't mean sloppy work; it means finding that sweet spot where a system functions reliably

without draining time and resources. After all, chasing perfection often results in spiraling costs, missed deadlines, and frustrated teams. If "good enough" keeps the system running, it's good enough for now.

For example, in software engineering, choosing a scalable architecture that addresses today's needs might be better than over-engineering for a future that may never come. The pragmatic approach ensures that systems are functional, scalable, and flexible without being overbuilt or unnecessarily complex.

However, there are high-effort and complicated practices that can significantly improve engineering outcomes: upfront multi-level modeling, requirements management and change tracing, system aspects alignment, up-to-date documentation, captured decisions log, and many others. Still, we rarely see them applied to real-world engineering projects. We used to think we needed a bigger engineering team and plenty of time to do high-quality engineering. The rise of lean and agile methodologies was a remediation for the inability to apply highly expensive practices on smaller and medium-sized projects.

AI's Shift: From "Good Enough" to Perfection—Affordably
Here's where AI steps in and upends the status quo. With AI, we can move beyond the pragmatism of "good enough" and start asking: "Why not perfect?" Traditionally, many engineering decisions were based on assumptions. AI can make more formal calculations; for example, with the proper prompt and specialized tooling, we can estimate how implementing additional features can contribute to the overall system load and evaluate potential remediations.

Balancing system qualities remains a constant challenge. Engineers frequently have to juggle factors like scalability versus cost, security versus performance, and reliability versus agility. Each of these dimensions comes with its trade-offs. Perfect security can slow down a system's performance; hyper-scalability can drive up costs. The trick lies in finding the right compromise, ensuring the system can handle current needs without overburdening resources. Again, AI can help us avoid assumptions and rely on more formal calculations.

Good practice for systems architecture design suggests making two to three design candidates to avoid cognitive biases. Now, with AI, we can prepare a two-digit number of design candidates, make thorough comparisons, and recommend the optimal one. AI provides an unprecedented opportunity for an engineer to finally arrive at a "continuous engineering" state, where even the most minor change gets through the full-scale change management process, touching multiple levels and aspects of the systems (like public-facing API, data schema, data transfer objects, domain events schema, etc.) and making tracks in numerous documentation and knowledge management systems—all without manual efforts.

Ultimately, AI can make the main trick affordable: seamlessly running "continuous engineering" cycles for a system without interruption for maintenance. Usually, such a setup and infrastructure configuration are not affordable to most engineering organizations. Microcontainers or serverless functions, blue-green gradual deployment, live health checks, and auto-acceptance are not easy to build. With AI, it all becomes affordable for a smaller engineering team.

In the end, engineers need to deliver functional, reliable systems, and the quickest path to success is usually through pragmatism—a "good enough" solution that works at the moment. But with AI, that pragmatism doesn't have to be the final word. AI's ability to continuously optimize and adjust means we can get closer to perfection without breaking the bank. And who knows—maybe in the future, engineers will finally stop saying "good enough" and start saying, "Yeah, this is perfect."

2. Precision, Expertise, and Responsibility

The technological landscape has always been dynamic, but it has become an outright avalanche in recent years. Every week, new frameworks, libraries, and platforms hit the market, each promising to be faster, more efficient, or more scalable than the last. This creates a monumental challenge for engineers—keeping up with constant change while still delivering stable, functional systems.

Traditionally, this meant continuous learning, constant experimentation, and staying on top of emerging trends. Engineers had to sift through release notes, compare features, and often invest significant time in trial and error. Adopting a new technology always comes with the risk of missing out on a superior alternative released a week later. The sheer volume of new information and tools is overwhelming, leaving many wondering, "How can we keep pace without burning out?"

Unfortunately, good engineering competencies are insufficient to build successful fit-for-purpose systems. After all, we expect engineers to improve and automate our transportation, healthcare, communication, energy, and many other industries. Consumer products are also impossible to make without business domain knowledge. For an engineer, it is a luxury to become an expert in more than one business domain, and only the most proactive can have three to five domains under their belt.

Fortunately, modern AI has agent-based architecture, which makes it an ideal solution to this problem. An AI agent is an LLM that has been trained in a specific area of expertise and equipped with appropriate tools, allowing it to perform all necessary actions to fulfill the possible domain-specific objectives. For example, we can have a "cloud security expert" agent equipped with a cloud infrastructure scanner, management console, and access to historical data (operational logs, analytics, modification traces, etc.). Alternatively, an "integration expert" agent can analyze external system API protocol, build a proxy connector, and provide the engineer with convenient code libraries that encapsulate the legacy stack and protect the system from compatibility and portability issues.

In general, you can build your "personal AI SWAT team," instantly connecting you to expertise in engineering and business domains. Instead of engineers needing to review and evaluate every new framework or tool manually, AI agents can take on the role of a "tech advisor." These AI-driven systems can continuously monitor the market for emerging technologies and analyze their relevance to ongoing engineering projects. Imagine an AI-powered assistant that reviews your existing architectural decisions and highlights areas where newer, potentially better alternatives have surfaced. The AI doesn't just propose changes blindly; it can contextualize its suggestions by comparing the pros and cons of different technologies, outlining trade-offs, and identifying risks associated with alternative architectures. It could provide a cost-benefit analysis, showcasing potential gains (like performance boosts or improved security) versus the risks (such as increased maintenance or the learning curve for developers).

AI agents can assess key potential benefits and hidden downsides of design candidates, providing engineers with a clear overview of the risks, for example, by addressing the following questions:

- Is the new technology mature enough to be reliable?

- What are the potential long-term maintenance costs associated with adopting it?

- Will this increase our technical debt by introducing new dependencies or incompatibilities with legacy systems?

- Is it scalable enough for our anticipated future needs?

The most developed AI agents will even be able to perform a migration to a new tech if needed, eliminating this kind of "technical debt" from the engineering backlog.

For business domains, an AI agent can perform a subject-matter expert role to define and prioritize usage scenarios, identify non-functional quality objectives (throughput, latency, load tolerance), formulate SLO range and key business metrics, etc. It can also be a source of "reference architecture blueprints" as a generalized and commonly accepted way of solving domain-specific problems and building domain tooling. This definitely helps a business to achieve an impressive quality baseline by avoiding functional gaps in a product and ensuring acceptable quality of the underlying systems.

AI empowers engineers to keep up with the rapid changes without falling behind or getting overwhelmed. It offers an adaptive approach to system architecture, ensuring that engineers keep up with change and stay ahead of it. This allows engineers to focus on what they do best: solving problems, creating resilient systems, and pushing the boundaries of innovation without constantly worrying about the following extensive framework or new business domain uncertainties.

3. Reflection and Bias Elimination

Far from just a tool or assistant, AI has the potential to reflect back on our decisions, thoughts, and processes, enabling deeper self-reflection and growth. It acts as a mirror, showing us where we are making sound decisions and where we might be veering off track, often without realizing it. This reflective capability is paired with an invaluable trait: eliminating cognitive biases and emotional influences that so frequently cloud human judgment. As engineers, the ability to reason logically and without bias is paramount. AI can be the partner that ensures our decisions are rooted in reality and aligned with the best possible outcome.

Every decision made in the design and architecture of a system has ripple effects that extend far beyond the immediate project. Whether it's the choice of technology stack, the approach to scalability, or the prioritization of features, each decision is critical to the long-term success of the system.

Traditionally, engineers relied on peer reviews, retrospectives, and post-mortems to assess their work. While these are effective to some extent, they often come with limitations—time constraints, subjective feedback, and sometimes a lack of depth in analysis. This is where AI's reflective capabilities come into play. AI can provide continuous, unbiased, and highly detailed feedback on engineers' decisions, allowing them to reflect on their work in real time.

For example, an AI system integrated into the development process could analyze code patterns and suggest improvements not just based on technical standards but on previous decisions the engineer has made. It could highlight areas of improvement, flag repeated mistakes, and recommend best practices derived from vast datasets of successful projects. This immediate feedback loop fosters a culture of continuous learning and self-improvement, allowing engineers to become more aware of their habits, strengths, and areas for development.

Moreover, AI doesn't just reflect on technical aspects. It can provide insights into workflow efficiency, communication patterns, and even team dynamics. For instance, it could alert an engineer that their tendency to over-communicate on specific tasks might slow down the team or that their coding style consistently introduces minor inefficiencies that accumulate over time.

Human beings are, by nature, prone to cognitive biases. These mental shortcuts or tendencies can lead us to make irrational or suboptimal decisions. For engineers, cognitive biases can manifest in several ways, such as favoring a particular technology because it's familiar (the "status quo bias") or being overly optimistic about project timelines (the "optimism bias"). The consequences of these biases can be significant—ranging from costly technical debt to missed deadlines and compromised system performance.

One of the most powerful applications of AI is its ability to detect and counteract cognitive biases. Unlike humans, AI doesn't have emotional attachments or preconceived notions about a given technology or decision. It might not always process information purely based on data, facts, and logic—but at least AI has "averaged" or "compensated" bias, making it an ideal partner in decision-making. Moreover, with minor effort, the AI system can report back to the engineer all the concerns related to potential bias, providing an opportunity to learn and improve their own cognition and decision-making process.

Let's take the example of engineering iteration planning. When estimating project timelines, engineers often fall prey to the "planning fallacy," where they underestimate the time required to complete tasks. AI, having access to historical data on similar projects, can provide more accurate time estimates, helping teams avoid unrealistic expectations and the subsequent rush to meet deadlines.

Another common bias is the "confirmation bias," where engineers might unconsciously seek information supporting their initial assumptions while ignoring data that contradicts them. AI, however, can flag these inconsistencies and present alternative viewpoints, encouraging engineers to consider all relevant information before making a final decision. This leads to more well-rounded and informed choices, ultimately improving the quality of the system being developed.

While emotion can drive creativity and innovation, in engineering, it can also cloud judgment, especially when stress or pressure is involved. Tight deadlines, budget constraints, and stakeholder demands can push engineers into making emotionally driven decisions, such as cutting corners on testing or prematurely deploying a system. These decisions may offer short-term relief but often result in long-term problems.

AI serves as a "neutral advisor," free from emotional influences. It assesses situations based solely on the available data and suggests rational and pragmatic actions. For instance, if an engineer is considering skipping a critical performance test to meet a deadline, the AI system could flag this decision, reminding the engineer of the potential risks involved—such as system crashes or performance bottlenecks under load.

Moreover, AI can help "de-escalate" high-pressure situations by providing calm, objective advice. In a crisis, such as a system outage, emotions can run high, leading to rash decisions that might exacerbate the problem. AI, however, remains unaffected by stress and can guide engineers through a logical, step-by-step approach to problem-solving. This ensures that the root cause of the issue is addressed systematically rather than applying quick fixes that might create additional technical debt.

One of the lesser-discussed benefits of AI is its ability to boost an engineer's self-confidence. By providing consistent, reliable feedback, AI helps engineers feel more secure in their decisions. Knowing that their choices have been reviewed and validated by an unbiased system allows engineers to proceed confidently, reducing the second-guessing and self-doubt that can sometimes plague complex projects. For example, if an engineer is unsure about the performance implications of a particular architectural decision, the AI system can run simulations and provide detailed feedback on the potential impact. This empowers the engineer to move forward with greater assurance, knowing their decision has been thoroughly vetted.

Additionally, AI's feedback loop helps engineers learn from their mistakes without fear of judgment. By continuously analyzing decisions and outcomes, AI can identify error patterns and suggest ways to avoid them in the future. This creates a safe space for engineers to experiment and grow, knowing that the AI is there to keep them on track.

In an ever-changing and increasingly complex engineering landscape, the role of AI goes far beyond simple automation or task management. It becomes a true partner—offering solid feedback, eliminating cognitive biases, and keeping emotions in check during decision-making. By doing so, AI enables engineers to make smarter, more informed choices that lead to better systems while fostering continuous learning and self-improvement.

About the Author

Ivan Padabed is the co-founder and CEO at system5.dev, a startup focused on systems architecture AI tooling; the director of platform architecture at Pandadoc.com (B2B SaaS unicorn); and an expert at Primary Venture Mastermind Network.

He is an experienced IT professional with 25-plus years in the industry, fostering great teams and building high-end products as a systems architect. Ivan is also a community leader, conference speaker, blogger (https://medium.com/system5-dev), systems engineering discipline evangelist, and author and instructor of multiple engineering and architecture courses. He's an active researcher in the field of systems architecture.

Ivan lives in sunny Portugal with his wife, four children, and a cat.

Email: ivan@padabed.com
Website: https://system5.dev/
LinkedIn: https://www.linkedin.com/in/ivanpadabed/

CHAPTER 23

THE AI REVOLUTION IN PERSONAL CYBERSECURITY: WORK SMARTER, LIVE BETTER

By Tanmay Patani, MS, GCFA, EnCE, ACAMS
Cybersecurity Leader, Speaker, Author, Advisor
Chicago, Illinois

> *AI can free us from tedious tasks, allowing us
> to focus on our creativity and innovation.*
> —Satya Nadella

Imagine waking up one morning to the alarming discovery that your bank account is empty, your personal conversations have been exposed, or, even worse, your digital identity is being used to commit crimes you have no knowledge of. This isn't the plot of a futuristic thriller—it's a very real danger in the age of AI-driven cyber threats.

We live in a world where digital life is becoming as integral to our existence as the air we breathe. Our data—our identity, our financial information, our private conversations—exists in the digital ether, subject to both the advancements of AI and the malicious efforts of cybercriminals who know how to exploit it. The issue? The same technologies meant to enhance our lives are often the same ones being used to compromise our security.

Our world is increasingly resembling the one in Steven Spielberg's 2002 film *Minority Report*, where technology anticipates your every move, for better or worse. The lines between real and artificial have blurred, and the potential for misuse is frightening. Just like the pre-crime unit in *Minority Report* stops crimes before they happen, AI today can predict, prevent, and protect us from cyber threats—or, if in the wrong hands, it can be used to exploit our vulnerabilities.

Experts predict that in 2025, 90% of online content will be generated by AI. Imagine navigating a world where almost everything you read, watch, or listen to has been synthetically created. In such a landscape, who can you trust? While the conveniences of technology—instant connectivity, AI-powered assistance—are undeniable, they also introduce new risks. In this chapter, we'll dive into both sides of AI in cybersecurity: the threats it poses and the opportunities it offers for us to work smarter and live better.

AI Threats: When *Minority Report* Becomes Reality

AI-driven cyber threats are no longer confined to the realm of science fiction. As AI evolves, so do the tactics used by cybercriminals, with an alarming ability to predict and manipulate human behavior. These AI threats can exploit vulnerabilities before you even realize they exist, creating an atmosphere of digital uncertainty.

One of the most insidious threats is deepfake technology, which uses AI to create synthetic media—videos, audio recordings, and images—that appear completely real, making it almost impossible to distinguish between what's real and what's fake. Consider the terrifying reality of AI-generated deepfakes. A multinational corporation fell victim to a $35 million fraud when a threat actor used an AI-created impersonation of the CFO to authorize fraudulent transactions. The technology was so convincing that even the company's finance team had no idea they were dealing with an imposter. It's a chilling reminder that AI is not just used to predict crimes in the future—it's being used to manipulate and deceive in the present. This is just one example of how the lines between reality and deception are blurring thanks to AI. As we continue to move into a future where our lives are mediated by AI technologies, we must understand the risks of these innovations, particularly when they can be weaponized.

Another chilling example is the rise of AI-driven identity theft, where fraudsters replicate a person's voice or image to carry out fraudulent schemes. In one case, a grandmother almost sent thousands of dollars to a cybercriminal posing as her grandson, using an AI-generated voice clone. This type of "pre-crime," where AI anticipates your emotional responses and manipulates you, is becoming all too common.

These examples highlight the dual nature of AI: while it offers incredible potential for protecting our digital lives, it also opens the door for unprecedented risks. The question becomes: how do we protect ourselves when the tools designed to keep us safe are also being weaponized?

Emerging AI Cyber Threats: The New Era of Preemptive Attacks

AI is a powerful tool that preempts attacks by studying human behavior, finding weaknesses, and exploiting them. Cybercriminals now use AI to plan and execute attacks before you even realize you're under threat. This is not science fiction—it's happening now, and it's affecting everyone. The following AI-driven attacks are becoming more sophisticated and harder to detect:

Confused Pilot Attack

Threat: imagine a world where AI systems you trust—like Microsoft 365 Copilot—are quietly being manipulated behind the scenes. Attackers introduce malicious content into the documents referenced by the AI, warping its responses and skewing decision-making. It's like tampering with the pre-crime system in *Minority Report* to wrongly accuse the innocent.

How to Defend: to defend against this type of attack, use AI-enhanced security tools that actively monitor for anomalies and warn you when your trusted systems are compromised.

Automated Phishing: The Perfect Lure

Threat: remember the personalized advertisements in *Minority Report* that scanned your retina and catered ads to your preferences? In a similarly intrusive way, AI-based phishing scams know you, too. By analyzing your behavior, threat actors can craft an email so personalized you may be convinced it is authentic.

How to Defend: employ AI-powered phishing detection systems that analyze your emails and links in real time, flagging suspicious communications and protecting you from scams before they even land in your inbox. These systems use advanced algorithms to identify even the smallest signs of fraud.

Malware Evolution: AI That Outruns You

Threat: In *Minority Report*, the technology that predicted crimes always seemed one step ahead of the pre-crime division. In our reality, malware is doing the same, evolving faster than we can build defenses. AI-driven malware adapts in real time, changing its attack vectors every time it encounters resistance.

How to Defend: AI-based antivirus and anti-malware software is the key to combating these adaptive threats. These tools learn from new malware attacks, identifying trends and modifying their defenses accordingly. They continually evolve, helping you stay ahead of ever-changing threats.

AI-Powered Social Engineering: Predicting Your Every Move

Threat: Much like the predictive technology in *Minority Report*, AI-powered social engineering knows how to manipulate you before you even realize it. Scammers use AI to create fake personas and hold highly convincing conversations, anticipating your reactions and playing on your emotions.

How to Defend: through data scraping, AI models learn what makes you tick. They analyze your conversations, your online presence, and even your vulnerabilities to build an attack that feels eerily personal. To defend against these attacks, use AI-driven security systems that help you monitor your online activity and identify suspicious interactions.

Deepfakes: Manipulating Reality

Threat: in the world of *Minority Report*, the idea that we can no longer trust our senses—our eyes, our ears—was central to the film's plot. Today,

AI-driven deepfakes bring that same fear into our real world. Whether it's forging a video to frame someone for a crime or creating a fake persona to manipulate someone into a financial transaction, deepfakes are the ultimate AI deception.

Advances in deep-learning technology have made it almost impossible to distinguish between real and fake. According to recent statistics, 90% of deepfake videos are designed to deceive, with 80% of those being used for malicious purposes like fraud, identity theft, or cyber blackmail. The potential is staggering. Deepfakes are being used increasingly in identity theft, with 58% of people being affected by such deceptive technologies. This represents a dramatic shift in how criminals exploit AI—making the tools of tomorrow part of the threats we face today.

How to Defend: employ AI-driven deepfake detection systems that scan media for inconsistencies and alert you to the possibility of fraud. These tools ensure that you can trust what you see and hear in an AI-dominated world. Deepfake detection is now an essential part of your personal cybersecurity toolkit.

Synthetic Media Fraud
Threat: AI can now generate synthetic media so convincingly that it feels like stepping into a manipulated reality. In *Minority Report*, the visuals and tech were breathtakingly real—and today, AI-driven fraud can make a false reality appear just as seamless.

How to Defend: with AI, cybercriminals can easily create synthetic identities, videos, and audio that mimic real people. This technology can be used for blackmail, fraud, or political manipulation, leaving victims trapped in a fabricated version of events. Use AI-driven verification services that cross-check identities and media across multiple platforms, preventing you from falling victim to impersonation or fraud.

Defending Yourself in a World Run by AI: Work Smarter, Live Better

While AI can be a weapon in the hands of cybercriminals, it's also your greatest ally when used wisely. Just like the pre-crime division in *Minority Report* prevented crimes before they happened, you too can leverage

AI to defend your digital life and prevent cyberattacks before they occur. The key is understanding how to harness these tools and integrate them into your daily routine. Here are some ways how you can use AI to work smarter and live better in an AI-driven world:

AI-Powered Antivirus and Anti-Phishing Tools
Use antivirus software that employs AI to detect and analyze threats in real time. These tools learn from previous attacks to continuously evolve and improve your protection. AI-powered anti-phishing tools can scan emails and messages, flagging suspicious ones before you even have a chance to click on a dangerous link. These systems provide a layer of protection that's continually updating, helping you stay ahead of new threats.

Automate Password Management with AI
AI-driven password managers create and store complex passwords for each of your accounts, ensuring that your credentials are secure. These tools monitor for breaches and immediately alert you if a password is compromised, allowing you to stay on top of your digital security without constantly worrying about forgotten passwords. With AI, you can generate passwords that are incredibly secure and nearly impossible for cybercriminals to crack.

AI-Based Identity Protection Services
Protect your identity with AI-based services that monitor the dark web for signs of misuse. These systems alert you if any of your personal information is being traded or used maliciously, giving you time to take action before it's too late. With AI keeping a watchful eye on your identity, you can sleep soundly knowing that you're actively defended against identity theft.

Monitor Your Digital Footprint with AI
Your digital footprint is the sum of everything you do online. AI tools can help track your online presence, alerting you if your personal data is being exposed or used without permission. These tools are designed to help you maintain control over your information, ensuring that you're not unknowingly putting yourself at risk by over-sharing or falling prey to data breaches. AI-driven security measures make it easier to protect your privacy and stay ahead of data leaks.

AI-Driven Behavioral Analysis for Proactive Security
AI systems can analyze your typical online activity and flag any unusual behavior. For example, if a threat actor gains access to your account, the AI system will likely detect suspicious activity, such as logging in from a new location or changing settings without your knowledge. These tools act as an extra layer of defense, helping you detect threats early on before any significant damage is done.

—

AI is a double-edged sword: while it offers cybercriminals the tools to manipulate and deceive, it also provides us with the technology to protect ourselves. In a world where cyber threats are evolving faster than we can keep up, the key to staying safe is embracing AI as both a shield and a proactive defense.

In *Minority Report*, the pre-crime system is ultimately shut down after it becomes clear that while it can predict crimes and prevent them, it also has significant flaws that can lead to innocent people being targeted. This highlights the double-edged nature of technology—what is designed to protect can also cause harm if not used responsibly. Similarly, AI in cybersecurity is a powerful tool that can be used for both good and evil. On one hand, cybercriminals can use AI to develop more sophisticated attacks, automate phishing schemes, and exploit vulnerabilities faster than ever before. On the other hand, AI provides robust defenses, such as advanced threat detection, real-time response mechanisms, and predictive analytics that can identify and mitigate potential threats before they become serious issues.

The key to leveraging AI effectively in cybersecurity is to recognize its potential pitfalls and use it responsibly. Embracing AI as both a shield and a proactive defense means constantly evolving our strategies and being vigilant about ethical considerations and safeguards to prevent misuse. By doing so, we can harness AI's capabilities to protect against the rapidly changing landscape of cyber threats. By leveraging AI to secure your digital life, you can work smarter—automating security measures, minimizing risks, and staying one step ahead of threats. More importantly, you can live better with peace of mind, knowing that you are actively defending your identity, privacy, and financial security.

In the end, the future of cybersecurity lies in how well you can adapt to this new AI-driven reality. The tools are available and technology is evolving, and by integrating them into your daily life, you can ensure that AI works for you, not against you. Embrace these intelligent tools and strategies, and navigate this AI-powered world with confidence, ensuring that your digital life remains secure. So, let's work smarter. Let's live better. And let's stay ahead of the curve in an AI-driven world.

About the Author

Tanmay Patani is an acknowledged and passionate cybersecurity leader with over 20 years of experience. His expertise spans cybersecurity strategy, insider threats, fraud investigations, incident response, threat intelligence, risk assessment, and compliance with a strong focus on leading digital forensics and incident response (DFIR), recovery, resiliency, eDiscovery, and litigation teams. He's collaborated with Fortune 1000 security teams and private equity firms to develop and implement enterprise-wide cybersecurity and data breach prevention strategies across diverse industries. He specializes in creating and executing tactical security programs that align with organizational goals, comply with regulations, and adhere to industry standards.

Tanmay holds a master's in computer science and a bachelor's in engineering, along with numerous cybersecurity certifications. A frequent speaker at industry cybersecurity events, he is recognized for his expertise and leadership.

Email: tanmayvip@gmail.com
LinkedIn: https://www.linkedin.com/in/tanmay2/

CHAPTER 24

AI: TRANSFORMING IT SUPPORT TODAY AND TOMORROW

By Greg Pellegrino
Vice President of Engineering
Cypress, Texas

> *I wanna heal, I wanna feel, like I'm close to something real.*
> —Linkin Park (from "Somewhere I Belong")

AI Today: Learning, Failing, and Growing

Artificial intelligence (AI) is currently driven by large language models (LLMs). LLMs work by predicting the next word in a sequence, with the prediction based on complex patterns learned from vast amounts of text. By training on diverse data, LLMs internalize relationships between words, grammar, facts, and context, which allows them to generate coherent and relevant responses. The chain of thought (CoT) technique is used to further enhance reasoning. Essentially, CoT prompts the model to create intermediate reasoning steps when solving complex problems, rather than directly outputting a final answer. This approach simulates human reasoning by processing it step by step.

While impressive, AI still needs to improve due to inherent limitations in how LLMs operate. The models lack true understanding and awareness. Responses are generated based on probabilities rather than real comprehension. They are prone to errors called hallucinations. Hallucinations with incorrect information can seem very credible due to their valid sentence structures. LLMs struggle with tasks that require deep contextual knowledge, common sense, or real-world experience.

Despite these flaws, AI is often "close enough" to be useful for many applications, like drafting content, answering questions, or summarizing information. Its ability to quickly generate relevant, coherent responses makes it a valuable tool for tasks that benefit from rapid output, as long as human oversight is used to ensure accuracy and reliability.

AI for IT: The Current Landscape

One use case that has made inroads is applying artificial intelligence to IT support. It's being applied to routine tasks, speeding up response times, and improving user experiences, changing how problems get solved. The promise is there, but the reality is sometimes messier. Chatbots can get confused by complex issues, automated ticketing systems can misclassify or escalate incorrectly, and predictive maintenance can generate false positives, leading to unnecessary interventions. The tools are impressive, but they aren't perfect yet, and it's important to acknowledge these current limitations.

Picture an IT department bogged down with endless tickets, routine queries, and repetitive tasks. That's where AI makes a difference. Chatbots and virtual assistants work tirelessly around the clock, handling the small stuff so that human experts can focus on what matters. Microsoft's Azure Bot Service or IBM Watson Assistant is there to answer questions, solve easy problems, and let humans take a breath.

And then there's the ticketing mess—manual entries, missed priorities, and endless backlogs. AI is here as well, automatically classifying support tickets based on urgency, content, and priority. It's like having a superpowered dispatcher who knows which problems need solving now and which can wait. ServiceNow's AI-driven platform is already doing this, making sure nothing falls through the cracks—but it's not flawless. Sometimes, important tickets get miscategorized, and while AI is growing, it can still make mistakes that cost time.

Think about predictive maintenance. AI is analyzing data from everywhere—network logs, diagnostics, and user patterns. It sees what's coming before it happens. Splunk uses this tech to predict when the subsequent failure will hit, letting teams fix things before they break and keeping everything running smoothly. However, these predictions aren't always spot on. False alarms can lead to downtime for equipment that was actually fine, frustrating users and wasting resources. For instance, a false alert might unnecessarily cause an entire server to be taken offline, disrupting business operations and causing delays for teams relying on that server's availability.

AI doesn't just predict; it solves. The common issues that suck up time—resetting accounts, updating software—it handles them without a second thought. Tools like Resolve Systems are doing this now, automating the simple stuff and leaving IT teams free to focus on more significant challenges like implementing strategic IT initiatives, improving system architecture, and enhancing cybersecurity measures. But even here, AI occasionally falls short—failing to correctly resolve more nuanced issues that need human intuition.

Knowledge management? AI's got that too. No more endless scrolling through outdated knowledge bases. AI looks at the patterns, figures out what people need, and makes it accessible. Tools like Freshservice are leading this change, ensuring that information is always up to date and easy to find. Yet, there's still the challenge of AI serving outdated or irrelevant information because it has yet to fully understand the context or changes in organizational needs.

Natural language processing (NLP) makes the experience feel human. It's not just text on a screen; it's conversational and intuitive. Google Dialogflow, for example, transforms how IT help desks respond, recognizing the nuances behind every question and providing answers that make sense. Still, NLP systems sometimes misunderstand complex or ambiguous queries, leading to frustration rather than resolution.

Workflow automation, user behavior analytics, sentiment analysis—AI is pushing IT to new heights. It's taking over the mundane, recognizing threats before they escalate, and providing insights into how users really feel. Companies like Darktrace use AI to monitor user behavior, catch anomalies, and keep systems secure. Platforms like Qualtrics gauge user feedback, helping IT teams get better every day. But these technologies are not infallible. AI can miss subtle indicators or flag false positives, leading to either missed risks or unnecessary alarms.

The Path to Smarter AI: Learning from Mistakes

But the potential is there. AI is improving rapidly, and with continuous advancements in training models, contextual understanding, and data integration, it's only getting better. As AI technology advances, we expect significant improvements in several areas. AI systems will be able to leverage even larger datasets to train on more nuanced scenarios, resulting in fewer errors and more accurate responses. Enhanced NLP capabilities will allow AI to comprehend more complex queries with greater contextual awareness, reducing misunderstandings and improving user interactions. Furthermore, continuous learning models will enable AI to adapt dynamically in real time, learning from its mistakes and refining its responses autonomously, providing a reassuring glimpse into the future of AI in IT support.

AI will also become increasingly proactive. Instead of just responding to issues, future AI will anticipate user needs and recommend actions before problems even occur. Imagine AI recognizing that an employee has a vital demonstration coming up and automatically optimizing their laptop for peak performance, or predicting a sudden spike in network traffic and allocating additional resources ahead of time to prevent slowdowns. Predictive maintenance will be fine-tuned to minimize false positives, and ticket classification will become almost flawless, ensuring that IT issues are addressed faster and more accurately. Collaboration between AI systems will improve, allowing multiple AIs to work together seamlessly, sharing data and insights to solve problems more efficiently.

These forward-thinking advancements will transform AI into a more powerful force capable of making a bigger impact on our daily lives. They will simplify complex tasks, enable smarter decision-making, and ultimately create a more efficient and resilient IT infrastructure. The hope is that these advancements will decrease false positives, make AI's decisions more reliable, and ensure fewer missteps, leading to smoother and more seamless operations.

The Future of AI for IT: The Assistant That Never Sleeps

Bright Horizons is a fictitious small company with big ambitions, and Iris, the AI assistant, is also a fictional creation. This scenario is intended to illustrate the future potential of AI through a practical example, showing how advanced AI capabilities could support and transform a small business. With a lean IT team and limited resources, Bright Horizons faces constant challenges—but they're determined to make it work. Bright Horizons is a small company with a bare-bones IT team, set three to five years from today. The world is chaotic, and AI is the one thing keeping the company afloat. The thing about AI is it's not just another tool. It's Iris, their AI assistant, and she's not here to make friends. She's here to keep things moving, no matter what.

Iris is the IT assistant who doesn't sleep, doesn't need a break, and doesn't care about feelings. Employees don't file tickets. They just tell Iris what the problem is, and Iris fixes it. She's not just a chatbot. She's a full-blown IT team member. She takes over devices, digs deep into issues, and doesn't stop until it's done. There's no "I'll get back to you." It's always now, always real time. Iris doesn't just stick to IT. When the sales team hits a roadblock in the middle of a client demo, Iris is there, pulling strings in the background—adjusting configurations, resolving software issues, and fixing connectivity problems—all without anyone noticing. She's the invisible fixer, the one making sure everything keeps going without a hitch. No stress, no disruptions, just smooth sailing. Productivity goes up, and stress goes down.

AI in the future is about knowing things before they happen. Iris watches everything—network traffic, devices, user behavior—and knows when things are about to break down. When the marketing team's laptops start showing signs of battery failure, Iris schedules maintenance before anyone even knows there's a problem. She's proactive and ruthless about it.

Security? Same deal. Iris watches the network, sees threats before they hit, and locks everything down. There's no hesitation, no asking for permission. It's automatic. Suspicious activity? Locked. Systems compromised? Shut down. Human experts only get the call when it's really needed. Bright Horizons is untouchable because Iris makes it that way.

Iris knows everyone. She knows Mary in project management needs her laptop optimized before presentations, and she makes it happen without being asked. It's personal. It's tailored. No one has to worry about their tools not being ready when they need them.

Iris isn't just about fixing problems. She teaches. If an employee keeps running into the same issue, Iris walks them through the solution. Over time, people learn, they get better, and they need Iris less. It's not just support; it's growth—a culture of getting stronger, smarter, and more capable.

Iris doesn't just keep systems running—she makes the business better. She analyzes data and gives insights into how IT resources are used, what's effective, and what's not. She helps Bright Horizons plan, invest in what matters and scale smartly. It's about value, not just uptime.

Iris doesn't wait for a human to intervene when something goes wrong with the cloud service. She talks directly to the cloud provider's AI, cross-references, and troubleshoots. It's AI talking to AI, fixing things faster than any human could. Downtime is minimized. Focus stays on the work, not on the broken parts.

Iris cares about productivity but also knows when people need a break. She optimizes schedules, keeps track of deadlines, and even tells employees when to step away for a while. It's not about overworking; it's about balance. Less stress, more focus, a better workplace.

Iris isn't here to replace humans. She's here to make them better. She frees up the IT team to focus on the big stuff—the creative strategic projects. She's the grunt worker, the one who takes care of the mundane, so humans can dream bigger. It's a partnership, and it works.

AI is already changing IT support, making it smarter, faster, and more intuitive. From automated ticket classification to predictive maintenance and intelligent workflow automation, AI is transforming how IT teams work—reducing repetitive tasks, improving response times, and proactively addressing potential issues before they become problems. In the future, with AI like Iris, the goal isn't just to save money or be more efficient. It's to make work better, make people happier, and create an environment where IT is seamless, invisible, but always there. Bright Horizons is just one small business, but with AI, it's ready to do big things. The future of IT isn't just about systems—it's about the people who use them, working smarter, living better, and not having to think about the tech because it just works.

About the Author

With over ten years of experience in engineering senior leadership roles, Greg Pellegrino currently leads engineering at Electric, a cloud-based IT platform. He is dedicated to delivering innovative and high-quality software solutions that address real-world challenges. Greg's expertise lies in leveraging AI and emerging technologies to enhance productivity, reduce technical risk, and improve the overall user experience. He believes in pushing the boundaries of what technology can do to create innovative, more efficient solutions. When he's not working on AI advancements, Greg enjoys strength training, outdoor and offroad adventures, and spending time with family and friends.

Email: roadkinggp@gmail.com
LinkedIn Profile: https://www.linkedin.com/in/gregjpellegrino/

CHAPTER 25

ARTIFICIAL CYBER WIZARDRY: THE INTERSECTION OF CYBERSECURITY AND AI

By Keith Pham, PhD
Cyber Risk and AI Leader
Fairfax, Virginia

The common media depiction of hackers as modern techno-wizards who—with a few glances at a glowing computer screen (as if interpreting mystical runes) and a blur of hands across a loud clacking keyboard (as if casting a spell—often complete with some muttered incantation)—can bypass any technological barrier and animate any machine to do their bidding is, of course, largely a work of fiction. However, the rapid development of AI technology and associated applications has been evoking fears of a different genre of fiction, specifically various forms of dystopian sci-fi in which humanity bends under the will of machines, whether through open might or with more subtle manipulation, or are replaced by the more advanced life forms.

While predicting the far future is famously an endeavor not generally known for accuracy, the current and near-future intersection of cybersecurity and AI poses plenty of challenges for which the world at large is not presently prepared. Of course, many efforts have been underway at various levels and institutions to develop safeguards to secure AI and leverage its cutting-edge advances in cybersecurity operations for both defensive and offensive purposes.

This chapter explores AI from a cybersecurity perspective, both in protecting AI in applications in the broadest sense and utilizing AI as a weapon, which is, itself, a topic understandably deserving of many chapters on its own. The entire, non-exhaustive exploration will only highlight high-level concepts and considerations and in a manner intended to be as accessible as possible for general audiences. Thankfully, many of the lessons learned and principles developed over decades in the cybersecurity discipline can also be applied towards AI.

The Dark Arts—AI in Cybersecurity—Offense

Discovering new vulnerabilities in software is a painstaking endeavor. It traditionally involves many hours of skilled experts poring over systems and applications and attempting to probe every possible entry point for any sign of a discernible weakness. Already, frontier LLMs have been utilized to automatically scan systems, applications, or networks for vulnerabilities faster and more efficiently than humans. Machine learning models can also identify exploitable patterns in software or hardware configurations. Essentially, the use of AI will lead to the rapid discovery of new vulnerabilities and exploits at a rate far faster than experienced before. Furthermore, AI already is being utilized to rapidly exploit those vulnerabilities as well.

Detecting an intrusion is also a task with which most organizations notoriously struggle. Roughly half of data breaches are discovered after a year and reported from outside the organization itself, often when valuable secrets are found in the darker corners of the internet. Detecting these intrusions will become significantly more challenging when conducted by AI-powered malicious software, which is already able to morph and mutate unassisted, completely bypassing defensive software, which is trained to expect specific attackers.

AI will also facilitate a new wave of attacks against what is considered the weakest link in cybersecurity—people. With images gathered (perhaps from social media) and a few snippets of voice, AI can create convincing voice and video replicas of people. It is incredibly common already for individuals to fall victim to phone, mail, and email scams. What happens when someone gets a video call showcasing a loved one bound to a chair, begging for their life and an immediate ransom is demanded? Such scams are also, sadly, already being conducted.

AI's potential in the offensive realm of cybersecurity is only just being scratched. Similarly, many efforts to leverage AI for defensive cybersecurity are also being explored.

Defense Against the Dark Arts—AI in Cybersecurity—Defense

One of the greatest issues in cybersecurity is the asymmetric advantage of an attacker versus a defender. A defender has to defend every vulnerability and entryway over time, never faltering. An attacker only has to find one entryway and one mistake. Modern applications consist of millions of lines of code and often a Frankenstein-like patchwork set of software sewn together. At a minimum, each "seam" presents the potential for a point of failure and vulnerability, which an attacker can exploit. Even if an application were to be perfect initially, the modern cycle of continuous improvement and upgrades will inevitably introduce errors into the process and offer avenues for a determined, malicious entity. This has been essentially a manpower issue that has always been advantageous to attackers, at least in the past. AI is changing that dynamic.

Capable, advanced AI is being deployed to shore up these boundaries, figuratively mending these seams and gaps, significantly reducing the avenues for attack. When a new vulnerability is announced by a software vendor, AI can be leveraged to block suspicious software based on behavior, instead of relying on specific signatures alone. AI can also be trained to download, patch, restart, and redeploy affected systems near-instantly. Furthermore, AI is also being used to analyze reams of organizational audit logs and network traffic to hunt for signs of intrusion around the clock. No longer will organizations be as reliant on teams of bleary-eyed analysts to manually stare at screens of data flashing by and expect them to connect the dots. With more advanced and capable AI, many of the asymmetric advantages that attackers have long enjoyed will shrink considerably.

Much more advanced techniques, which we'll only touch upon here, involve the implementation of moving target defense and adaptive cyber defense concepts, which constantly change the organization and system environments to stymie attackers, like the trope of a shifting labyrinth. Determined attackers usually require significant time and effort to conduct reconnaissance and learn their target and its environment, carefully crafting the attacks for the specific target. An environment that constantly shifts around them forces the attacker to reorient and renavigate, thwarting them at every turn.

While these more exotic solutions get a significant amount of attention, much can be learned from the lessons of the past that we can apply to our present.

Elder Knowledge—Lessons from Real-World Risk Management

Risk management as a professional practice rests upon a foundation of many fundamental principles. One such principle is the hard fact that fully eliminating risk is simply not realistically possible. This can be visualized in a non-technical analogy—securing a home. A person can install ten locks on the front door, but an attacker (in theory) can always pick each lock or (more realistically) bypass them altogether by using a window. Meanwhile, the day-to-day hassle of locking and unlocking all of these locks would cause such a burden that a family living in such a home would consider using the window themselves!

This highlights another principle: safeguards and protections generally incur a cost, whether to implement initially (e.g., buying and installing locks) or to maintain and operate over time. Thus, safeguards are only worthwhile when considering the risk involved, the costs (both initial and ongoing), and the effectiveness against the threats the safeguard is intended to counter. To illustrate, let's pull an example that is sci-fi-adjacent to AI. In a hypothetical situation where a highly advanced alien civilization suddenly appears from the vast reaches of space and attacks the planet, no realistic, effective safeguard exists. Such a civilization—if it were to ever appear—would be almost certainly so far advanced beyond humanity's technological capabilities (at least currently) that no defense is realistically possible. Thus, implementing any so-called protection specifically against this threat is realistically a waste of resources. The situation is similar to many other "end-of-the-world" or apocalypse scenarios.

Another common example of risk management mismatch found on planet Earth involves whole life insurance. In theory, this type of insurance provides for a spouse and dependents (particularly one's children) in the event of untimely demise or injury and covers one's entire life. However, the premiums for this type of insurance are incredibly costly over the long run. Even if the terms of the insurance policy are met, beneficiaries tend to lose significantly more money in comparison to a scenario where the exact same funds were instead banked, let alone invested. The insurance companies certainly make no secret that they're using the payments to make their own investments to increase profits. Furthermore, the individual or survivors often face complications due to the insurance claims submission process, risking the claim being delayed or denied outright when needed the most. The choice for whole life insurance is particularly excessive when someone does not have a dependent spouse or children. Even for those with children, purchasing term life insurance (with coverage extending until the children reach adulthood) is more appropriate for most.

In contrast to these examples, there is also no shortage of stories about organizations that neglected to prioritize risk management and cybersecurity, or focused too heavily on a single aspect. For instance, some overemphasize the prevention of intrusions and attacks while a sound defensive strategy should also consider cost-effective measures to reduce impacts if an attack does occur.

In a nutshell, the most critical lesson from the history of risk management from the real world is that a clear vision of priorities—the possible protections and their potential pitfalls—is critical to minimizing risk overall. These lessons also apply to cybersecurity as well as the protection of AI-enabled services and applications.

Self-Scrying—Knowing Oneself

The starting point in protecting AI is very much the same as protecting other applications: asking questions. For example: what does it do? What value does it bring? What's the impact if its data gets leaked, corrupted, or the application as a whole goes down? If it's down or otherwise in a reduced state, how long can it remain in that state before the impact is unacceptable? In the case of critical business applications, is that downtime perhaps a threat to the organization itself?

A simple news aggregator application that retrieves data from publicly-available sources should be secured very differently than a business intelligence application that crunches proprietary customer data for insights or a defense fusion application that combines intelligence from multiple classified databases to provide tactical insights. To be more explicit, the public data gathered in the news aggregator is not sensitive, so it shouldn't be protected as if it's a military-grade secret like the defense fusion application. The business intelligence application is clearly somewhere in between.

Traditionally, ensuring that the application remains operational is often more important than absolute secrecy in many scenarios. If an ongoing military operation is reliant on the defense fusion application to provide real-time situational awareness, downtime could result in the failure of the entire operation and many lives lost. By comparison, if the secrecy of the communications is broken later, the operation may have been long in the past and the value of that information would likely be minimal. Similarly, it's acknowledged that business downtime during peak holiday season is also often extremely costly and insider trading generally isn't an issue once company information becomes publicly known. The time value of information is critical.

This is far from dismissing the importance of ensuring the secrecy and privacy of data, especially in AI applications. A key differentiator versus the applications of yesteryear is the sheer scale of the datasets required and processed for AI. This data is necessary to identify patterns, make predictions, generate content, and provide all of the other benefits that AI offers. However, the massive volume of raw data also presents many concerns. Among them is the ability to glean far deeper organizational and individual insights than could be achieved previously, some intensely personal. Stories abound of AI-powered stores recommending pregnancy-related products to individuals before they even realized they were pregnant. It is, therefore, more important than ever to ensure data is handled appropriately, in accordance with established cybersecurity recommendations.

Mystical Manuscripts and Tomes—Standards Organizations, Guidance, and Legal

Leading cybersecurity organizations are placing greater emphasis on developing AI-specific standards and frameworks. Similarly, organizations not traditionally centered on cybersecurity, such as those specializing in data science, analytics, and data privacy, are increasingly providing cybersecurity-related contributions. This highlights the expansive reach of

AI and its intersection with numerous domains. These collective efforts offer valuable guidance for mitigating risks, promoting ethical practices, and protecting AI systems against adversarial threats. Adopting these recommendations, tailored to specific industries, is prudent—even beyond the context of AI.

The following is a non-exhaustive list of these organizations and notable guidance, in no particular order:

US National Institute of Standards and Technology (NIST)—AI Risk Management Framework

- International Standards Organization (ISO)—ISO 42001 AI Management Systems

- MITRE—Adversarial Threat Landscape for Artificial Intelligence Systems (ATLAS)

- Google—Secure AI Framework (SAIF)

- Databricks—Databricks AI Security Framework (DASF)

- Open Web Application Security Project (OWASP)—AI Security Exchange and Top 10

- International Association of Privacy Professionals (IAPP)—Artificial Intelligence Governance Professional (AIGP) Body of Knowledge (BoK)

Following the guidance provided by these organizations at a minimum also demonstrates a concern for due diligence and assists in compliance with AI legal frameworks being developed worldwide, with the EU AI Act in 2024 as one of the most prominent recent examples. This, itself, is a vast topic best addressed elsewhere.

Zen Minimalism—Reducing the Attack Surface

Once one has a firm understanding of the AI-related system or application as well as the legal frameworks in which it operates, the first actions are often to limit functions and access that don't provide value or are unaligned with the purpose/mission. Such functionality presents an opening for an attacker. Unfortunately, removing existing features and components can entail significant, time-consuming changes. This is one of the reasons security should be involved early in the application ideation and design phases to reduce potential rework and the associated expense and lost time.

Similarly, user access to the application and functions must be examined and limited to only those permissions necessary to complete their purpose or role. AI applications and systems bear similarities to microservice architectures, which can involve complex interactions of individually-limited components that combine to produce almost "magical" results.

A common use of AI agents involves a multitude of them independently communicating with each other to interpret natural language input, determine appropriate sources of information, negotiate its retrieval, and then analyze and contextualize the response for the user. Each interaction presents a security concern. Cataloging the interactions and imposing reasonable limitations is the goal. Any AI associated with retrieving public news aggregators should never be requesting information from AI associated with the classified defense fusion application.

Pacts and Contracts—Cost Considerations

Despite earlier examples, many of the exercises that mature organizations undertake to improve their security posture can lead to considerable savings in the long run. Systems, applications, and functionality, which aren't aligned to the organization's purpose and mission, are often deprioritized and have their resources allocated elsewhere right before a security review. Removing unnecessary connections and access results in less traffic, the removal of unnecessary software triggers a review of software inventories along with associated software licensing, and the removal of unnecessary user access removes or minimizes the need for a host of user support features. AI functionality often requires prohibitively expensive and specialized hardware. This hardware has become so critical, nation-states have begun investing vast sums to invest in their domestic manufacturers to secure national security interests. Proper AI governance can paradoxically lead to massive cost savings and performance improvements in addition to the intended goal of promoting security.

Curtain Call—In Conclusion

As famed science fiction writer Arthur C. Clarke once said, "Any sufficiently advanced technology is indistinguishable from magic." Currently emerging AI technology—while one day quaint from the perspective of the far future—is inarguably the stuff of supernatural wizardry compared to what has come before. The potential of AI is both at times otherworldly and in many cases unnerving. Make no mistake: even though we can lean on the lessons of the past, protecting the systems, services, data, and people that enable AI technology from a cybersecurity perspective presents new challenges, as well as twists on old ones. AI offers power and promise for both cyber attack and defense. At the beginning of this chapter, we mentioned that predicting the future is notoriously difficult. However, it is clear one prediction is quite easy: AI will be transformative, both in cybersecurity and in society as we know it.

About the Author

Keith Pham's career journey has traversed border stations, aerospace, the Pentagon, and the US Intelligence Community. He has served the Big Four, MBB, and the financial sector in a variety of cyber roles, managing multiple security programs to include establishing strategic direction, overseeing implementation, and evolving the programs over time to contend with the shifting cybersecurity landscape, including implementation of artificial intelligence (AI) cyber risk management.

Keith holds an Information Security Ph.D. with an academic focus in deploying Adaptive Cyber Defenses (ACD) to mitigate advanced persistent threats (APTs). He plays tennis and is a classically-trained foil fencer.

LinkedIn: https://www.linkedin.com/in/keithpham/

CHAPTER 26

YOUR BOT, MEET MY BOT: HOW AI IS TRANSFORMING RECRUITMENT

By Nikki Remkes
Founder, Remkes People Solutions, HR & AI Strategy
Heerewaarden, The Netherlands

> *Open your arms to change, but don't let go of your values.*
> —Dalai Lama

Back in 2019 I first got acquainted with (conversational) artificial intelligence (AI) working as a talent acquisition manager for a global corporation. During this time, most thought AI meant "ad interim," so they related it to freelance work and were not very open to it. Apparently, they did not realize we were actually talking about technology.

I recognized AI might have potential in the many recruitment processes for the variated job seeker audiences I was managing at the time. We used the technology to enable job seekers to apply through WhatsApp for cleaning jobs. These job seekers often did not have CVs, motivation letters, or a computer, but they *did* have a smartphone. Job seekers were asked some simple-level questions about location and availability, for example. If there was a match with one of our vacancies, the person was invited for the interview. There was only one interview needed, as all the basics were already covered. Although it took some adjusting, the technology greatly contributed to our hiring goals, resulting in happy job seekers and hiring managers.

Today different types of AI technology are rapidly transforming the way we live and work. In this chapter I will focus on the main transformations within recruitment with the aim of providing some perspectives and guidelines both to recruitment professionals and job seekers on how to utilize AI as well as remain critical about its use. I believe AI is not here to take your job. However, it is essential to learn how to use it properly as it does have the potential to let you focus on what you enjoy most within your profession. Globally it could be an answer to do more with less people and act as a partial answer to the retirement crisis we see rapidly developing within the US and Europe.

Our World

Today our brains process incredible amounts of information on a daily basis. Technology is enabling and often stimulating us to do so. Just think of the functionality of the mobile phone, developed in the early '80s, adopted by the public in the '90s and early 2000s. Initially we were happy to be able to make a call, and today a mobile phone is a mobile computer in itself, providing us with access to numerous applications, which are (mostly) designed with the purpose to enrich our lives. We rarely use the mobile phone for its initial purpose though, actually making a phone call. We are even working with a multitude of technologies most users do not fully understand yet: "How *does* the app that I have just installed work? How does it learn and what happens exactly with my data I entrust to it?"

The same applies to organizations and employees who use AI in their recruitment process. Often the users do not have a full understanding of the AI they are deploying themselves. This can be viewed as strange and risky at times, as it can greatly impact your reputation as an employer and the overall success of your company. This is true especially in case your deployed AI negatively impacts your ability to attract talent. In a shrinking market for talent, this can have severe consequences.

Most talent attraction and people leaders I speak to strive for more quality and less quantity in every part of the employee journey. They often say, "True human attention is what matters." Sadly the labor market is trending in the opposite direction in both the US and also in Europe. We are moving towards a quantity market, where, on average, a job seeker needs to do over 200, sometimes close to 300, applications in order to schedule sufficient enough interviews to actually land a job they hopefully will enjoy.

A Day in the Life of a Job Seeker

Imagine, you don't feel the same motivation for your job anymore. This is what your gut feeling is telling you every Sunday evening, and it is not going away. Or the company is downsizing and you are one of the employees that is let go. Besides the emotional process you go through, you have a life to enjoy, a family to feed, and other responsibilities. You realize that work does provide meaning and satisfaction in life besides the practical aspects of it, so you start to look for a new job.

As you start looking, you are entering an incredibly laborious process of finding and applying for a new job, as it is fair to recognize that finding a new job is, in fact, a full-time job on its own. It is highly emotional, as you will probably often deal with unclear rejection if you hear anything at all after you submit your job application. You have read and heard about the revolution of AI. And it would be great if AI could take these mental blows for you and save you time as well in the application process. After using some search engines, you see that what you are looking for exists, in abundance, and you, hopefully carefully, select a competent AI tool to help you.

Often initially you are selecting the free version, as you want to test the technology first. You start feeding your selected AI tool with information—personal information from your CV such as your contact details, your address, and experience—but also you feed it with your hopes and dreams for a new occupation. Some of these might be straightforward such as base pay, hours, and your commute. But what you really think is important is more nebulous and more complicated to grasp, even for a human, let alone technology.

One of the most complex but detrimental aspects to ensure you are happy in your job is company culture. What is company culture and what should it encompass for you as a future employee? Most likely, also due to time constraints you provide your AI tool of choice the information—"great company culture"—and you are off to the job application races. I can only imagine your AI bot asking the employer's bot about the "great company culture" with the AI company bot answering: "We have a great company culture: we have fruit in the office, a foosball table, and deep fried Friday with company drinks. We, therefore, must be a match made in heaven. Let's schedule you for an interview with us. What is your availability?"

A Day in the Life of a Corporate Recruiter

Recruitment professionals have seen the recruitment landscape significantly transform this year, at an even more rapid pace than usual. Recruitment is in the never-ending and ever-changing pursuit to attract human talent. However, human behavior, especially in terms of which type of tech, social media, or even secondary salary benefits are desirable, can change in a day, as it's all subject to the next trend. Talent is spread out across different generations with sometimes very different desires, and it is your job to attract talented people to join your employer. Your day consists of highs following lows in quick succession, which requires a strong mental resilience.

The ideal vacancy ratio—how many vacancies a recruiter ideally should handle at the same time—is often discussed, but rarely executed. What this means is that, in practice, the recruiter rarely has more than an hour to spend per week per vacancy, should you be able to divide your time evenly. Next to the job seekers, who tend to ghost you more often, as you have recently noticed, you have an organization and its many internal stakeholders to manage. A very challenging job indeed! Fortunately, tech, in general, and recently AI, in particular, has been progressing rapidly to enable you to do more with less.

Of course, AI provides great functionality for enhanced efficiency, such as administrative tasks like writing job descriptions and summarizing your candidate screening or job intake calls. It can screen large quantities of applications for vacancies that have straightforward requirements, such as availability and location. This helps to accelerate the hiring process, as it can, for example, auto schedule interviews that fit the calendars of all participants.

Often AI technology claims to be reducing bias, and in a sense, this is true. It can check job descriptions for you on biases, for example. If you use it to screen potential candidates and, therefore, use it within your decision-making process, it already becomes more challenging. AI, as a successful technology, depends on the data you put in, how it is designed to learn, and the user needs to understand this. Often companies want to hire more A-players or top performers. But if your top performer is also not the nicest colleague to work with, you just might end up with a team full of them. Is this what you want to achieve? In some cases, yes; in most cases, definitely not.

The same applies to being more "data-driven," one of the trending words in the strategy of businesses across the globe in recent years. And this is also true, you do have more data points, hence more information to analyze, which either helps you in making a well-informed decision or creates an information overflow, which makes making the decision even harder. Of course, AI can get information about who historically has been successful in that specific role; however, most companies do not have any data on that to feed the AI. Most roles have been subject to change, especially in the past decade. When the data that is put in is not carefully updated, the AI bases its decision on old criteria, which can be totally fine given the requirements of the position. However, when we are dealing with new roles or jobs, with new responsibilities and skills being demanded on an almost daily basis, we can't rely on old data to make the right selection.

However, AI should be part of your toolstack or your equipment to be a successful recruiter. As a professional you have the responsibility to understand what it does and how it works. Suppliers of AI-powered assessments and screening tools are recognizing that the decision should not solely lie within the technology; there is an ethical and moral responsibility to audit the AI you use. Fortunately, we have seen another pivotal role on the rise in corporations, the role of data protection officer, who can help the recruiter in mitigating data privacy concerns and access to data and help to monitor if it is responsibly stored.

Which Way Forward Do We Choose?

Both job seekers and employers have a responsibility to understand and properly manage the respective AI they choose to use. Ask yourself, what would benefit you the most? Do you, as a job seeker, want to feed all your personal information to AI in order to send out 200 applications and land you 20 interviews? As an employer, do you want your candidate to chat to your bot first, and do you trust the bot to make the actual hiring decision for you?

Yes, we are biased. Yes, we do make mistakes when hiring people. We need to accept that humans are often messy and noisy while in the process of making a decision, as this is the reality we are living and operating in on a daily basis. Even still, I urge you to let your tech be supportive of human decisions by creating less noise, not more. Moreover, do not let AI take responsibility from you. You, as a job seeker, are responsible to find a job that fits you, as you are the one that needs to actually do the job on a daily basis, not your selected AI tool. As an employer, you choose to be a leader, to take responsibility and enable a job seeker to become an employee and to develop, grow, and be productive for your company.

As a job seeker, you should look for a job that makes you happy, as we tend to work for so many hours of our lives. So, it had better be good and meaningful as well. So, ask yourself, do you want two bots lying to each other or an actual human conversation with less noise in order to determine if you like the other person enough to take the commitment to the next level and become employee and employer? Zeynep Tufekci already mentioned it in her 2016 TED talk, when she explained that machine intelligence is here to stay, and, therefore, our human morals and values are more important than ever. Be curious, and understand what you use and why you use it.

About the Author

Nikki Remkes is the founder of Remkes People Solutions, a consultancy dedicated to enhancing organizational effectiveness through strategic human resource management and supporting (AI) technology. With a wealth of experience in recruitment, talent management, and HR strategy, Nikki has established himself as a trusted strategic advisor for businesses ranging from start- and scale-ups to large enterprises, seeking to optimize their workforce.

Additionally, Nikki is an associate of smartrecruitment.ai and a senior advisor HR strategy and recruitment at Ondernemers Adviseurs (business advisors) and works with both Dutch and international clients.

Email: nikki@remkespeoplesolutions.nl
Website: www.remkespeoplesolutions.eu

CHAPTER 27

UNLOCKING HUMAN POTENTIAL WITH AI

By Gabriel Lars Sabadin, PhD
ML & AI Engineer, Award-Winning Technologist
Stockholm, Sweden

> *The real problem is not whether machines*
> *think but whether men do.*
> —B.F. Skinner

Artificial intelligence (AI) is one of the most transformative technologies of our time, reshaping our world in ways that seemed unimaginable only a few years ago. Its disruptive potential reaches nearly every corner of our lives, altering how we work, play, and think. But for all its power, AI is not here to replace us—it's here to elevate us. This chapter explores how we can leverage AI to unlock human potential, navigate the complexities of this change, and ensure that our journey with AI remains positive. Here's how.

AI as an Augmenter, Not a Replacer

AI raises fears of replacement, but if, instead, we reframe it as an augmenter, it becomes a powerful ally—a tool that extends our abilities. AI can handle repetitive tasks, freeing us to engage in more creative, strategic, and fulfilling work. Imagine an artist using AI to spark fresh ideas or a scientist leveraging AI for faster data analysis. In each case, AI assists in enhancing human potential rather than diminishing it.

Our greatest opportunity lies in embracing AI as a co-pilot. We can integrate its strengths into our workflows to achieve more than we could alone. When we see AI as an augmenter, it inspires progress, empowering us to do our best work.

Ethical AI Development

With AI's growing influence, we must not overlook its potential risks and responsibilities. As AI continues to grow, ethical considerations are crucial. If left unchecked, AI has the potential to perpetuate biases, compromise privacy, and foster mistrust. My work in AI/ML has shown me the importance of developing AI responsibly, with clear ethical standards in place.

Consider a predictive algorithm in a hiring system: if not designed with fairness in mind, it may favor certain demographics over others. Ethical AI development requires careful thought—from fostering diversity in training data to implementing processes that prioritize privacy and transparency. We can ensure that AI remains a force for good by emphasizing fairness, respect, and trust in its design. When AI is ethically sound, it gains public trust and long-term sustainability. As developers, businesses, and users, we hold the responsibility to build and use AI in ways that benefit everyone.

AI in Everyday Life

AI is already embedded in countless aspects of daily life, often in ways we might overlook. It boosts productivity, enriches customer experiences, and provides convenience. Take, for example, a virtual assistant that manages your schedule, recommends stress-reducing activities, or assists with wellness goals. AI's role in healthcare is also transformative, aiding doctors in analyzing complex data and predicting treatment outcomes precisely.

Imagine a scenario where a smart assistant not only organizes your calendar but also suggests breaks based on your stress level and personal well-being. Or think of AI's impact on education, where adaptive learning platforms cater to individual student needs. These examples illustrate how AI is not only a convenience but a powerful ally, enhancing our lives in ways that are often subtle but significant.

Surviving AI Disruption

The rapid pace of AI advancement is both exciting and uncertain. As AI reshapes industries and automates tasks, we must adapt to remain relevant. Surviving AI disruption demands proactive strategies—up-skilling, exploring new roles, and embracing lifelong learning.

For those in evolving fields, building skills that complement AI, such as critical thinking, problem-solving, creativity, and empathy, is essential. While AI can process vast amounts of data and execute tasks efficiently, uniquely human qualities—like intuition, adaptability, and emotional intelligence—remain irreplaceable. Workers who embrace these strengths and stay adaptable turn potential threats into opportunities. Through resilience and continuous learning, we ensure that AI disruption becomes a force that empowers us rather than a wave that leaves us behind.

My advice to those navigating these changes is to identify the unique human skills that AI can't replicate and cultivate them. By focusing on what makes us uniquely human, we stay indispensable in an AI-driven world.

Navigating the Human-AI Relationship

Our relationship with AI is changing. Once a tool, AI is now a collaborator, fundamentally altering how we interact with technology and one another. The more we work alongside AI systems, the less we see them as novelties and the more they become integral parts of our lives.

In this new reality, understanding AI's capabilities and limitations is key. With this knowledge, we set realistic expectations and collaborate more effectively, achieving a balanced, productive partnership with AI. The future holds a world where humans and AI complement each other, each contributing unique strengths.

Picture an AI-enabled project where human creativity meets machine efficiency. By working together, humans and AI can push boundaries that neither could achieve alone. The future is collaborative, and the more we embrace AI as an ally, the further we can go together.

As we move forward, our challenge—and opportunity—is to shape AI as a foundation for human flourishing. This journey is not just about building smarter machines but about enhancing what it means to be human. Together, we can create a world where AI serves as a bridge to a brighter, more empowered future for everyone.

About the Author

Gabriel Lars Sabadin is a seasoned ML/AI engineer with a wealth of experience designing human-centered AI solutions for companies like Adobe, Apple, Porsche, Electrolux, Lufthansa, Citibank, and MasterCard. An MIT-trained PhD in computer science, he has developed custom LLMs for platforms such as LifePal, Baibe, Worthfit, PetPal, OrizAI, and SongFox. Gabriel is driven by a vision of accessible technology that enhances everyday life.

He lives in Sweden with his wife and their three children, who inspire his commitment to advancing AI responsibly. Outside of work, Gabriel is an avid collector of rare vintage guitars, valuing the history and artistry behind each piece.

Email: gabriel.lars.sabadin@gmail.com

CHAPTER 28

THE IMPACT OF GENERATIVE AI ON HIGHER EDUCATION

By Cedric De Schaut, MSc
Creator of Generative AI Bootcamps; Lecturer
Digital Nomad (from Brussels, Belgium)

*Education is the passport to the future, for tomorrow belongs
to those who prepare for it today.*
—Malcolm X

The transformative potential of education has long been recognized. Yet, it has also always been a hotly debated topic. Depending on which political party is governing, curricula might change drastically. Unfortunately, the 20th century in particular witnessed a troubling escalation of violence and loss of life in the pursuit of educational rights and reforms. Student movements, driven by a desire to challenge the status quo and shape their own learning experiences, often resorted to confrontational tactics against governments and universities.

Higher education institutions, once regarded as bastions of integrity and intellectual rigor, have faced significant scrutiny in recent times, leading to a degree of erosion of their infallibility. Scandals involving academic misconduct, financial mismanagement, and failure to address issues such as harassment and discrimination have tarnished their reputations.

Furthermore, there are two fundamental issues. Firstly, the financial burden associated with five years of higher education is a significant concern. Obtaining a bachelor and a master's degree in the US could easily set you back close to half a million USD, assuming you're someone with an average annual living expense of 10,000 to 18,000 USD. To obtain a positive return on investment (ROI), future earnings need to be pretty high. Tuition in Europe is generally much lower, but the overall investment is still considerable. In addition, many recent graduates are not fully prepared to enter the job market after investing five years in their studies. This lack of readiness is partly due to universities not prioritizing job-readiness as a primary objective. For instance, law or marketing graduates may lack essential skills, such as creating pivot tables in spreadsheet software, which are crucial for many entry-level positions. Moreover, universities often struggle to keep up with rapidly evolving technologies, resulting in outdated curricula and teaching methods.

Finally, an often-overlooked argument is that many recent graduates in the Western world are still uncertain about their career paths at the point of graduation. As a result, they may find themselves needing to pursue additional degrees or even additional master's programs to figure out their career goals or to get job-ready.

Short, intensive courses—hereafter referred to as bootcamps—offer a compelling alternative. Bootcamps solve many problems that universities are facing. Bootcamps are designed to be nimble and responsive to the rapidly changing job market. They are cheap and short compared to university degrees, there is less administrative burden involved, and they focus on skills rather than on prestige. Combining the advantages of the bootcamp model with the advent of generative AI can significantly boost productivity, democratize knowledge at a fraction of the cost of traditional institutions, and ultimately benefit society as a whole.

Generative AI and Multi-Agent AI Systems

Generative AI is a type of artificial intelligence that can create new content, such as images, text, audio, or video, by using patterns learned from existing data. As a GenAI content creator and instructor, I have been on the frontlines witnessing the impact of generative AI on education.

One of the core generative AI technologies poised to revolutionize education is the concept of multi-agent AI systems. An agent can be defined as a large language model (LLM), such as ChatGPT, equipped with cognitive capabilities such as memory and reasoning. Additionally, these agents can communicate with other agents, browse the web, and execute tasks, all autonomously. A multi-agent AI system consists of various agents with different skill sets, knowledge, and roles that interact with each other.

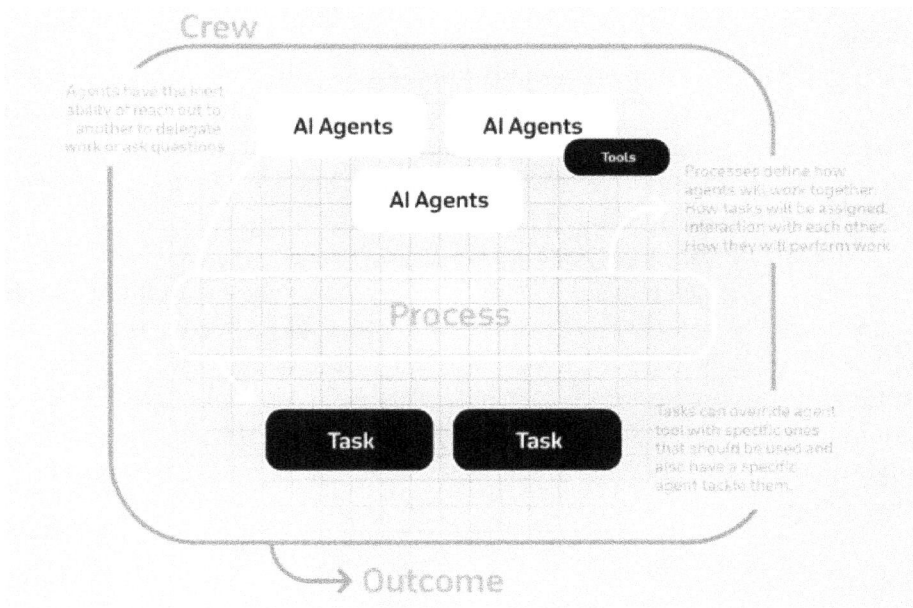

Fig 1.1: visualization of a multi-agent AI system of a popular framework "crew.ai"

A notable example is the ChatDev repository, which functions as a virtual software company operated by various intelligent agents each assigned different roles such as chief executive officer, chief product officer, programmer, reviewer, tester, and art designer. These agents collaborate within a multi-agent organizational structure. Agents within ChatDev collaborate by participating in specialized functional seminars, focusing on tasks such as designing, coding, testing, and documenting. This collaborative approach streamlines the software development process and enhances efficiency.

The Impact of Generative AI on Education

Generative AI is already significantly impacting higher education, offering two key benefits that stand out: content creation and content delivery. Currently, creating content for short, intensive training programs is a labor-intensive process that involves multiple steps, including curriculum development and the production of the actual training materials.

Curriculum Development

What

Curriculum development requires extensive research and planning. Instructors must identify learning objectives, structure the course, and ensure that it aligns with industry standards and job market demands. However, the time and cost pressures faced by curriculum designers often lead to suboptimal results.

Ideally, input from additional stakeholders such as pedagogical and subject matter experts is sought to enhance the curriculum development as well as the content creation process. However, this can be challenging, as contributors may not be compensated for their time, resulting in suboptimal results. Moreover, coordinating schedules, aligning content, and ensuring consistency can complicate the development of high-quality, streamlined content as well causing significant delays in the overall operation.

Impact

A multi-agent AI system can significantly enhance and expedite the curriculum development process. One agent could focus on researching specific topics by browsing through the web, academic articles, entire e-books, or even podcasts. This agent generates comprehensive summaries that are accessible to both other agents and human stakeholders. At this stage, an optional human check could be inserted to identify the need for correcting or adding information. However, some agents specializing in market research have already demonstrated superior performance in terms of research quality. The advantages of agentic market research become even more pronounced when factoring in the time and cost efficiencies that AI systems offer. Therefore, there is no essential need for human intervention or assistance in the process.

The next agent in the system can propose a potential outline for the course and submit it for validation to another agent responsible for assessing industry standards and job market demands. Once this validation agent approves the proposed structure, a notification could be sent to a designated human reviewer, containing the proposed curriculum and a brief explanation of how it addresses a gap in the labor market.

Upon receiving the green light from the human reviewer, the AI agent will finalize the process. This may involve creating a professional, internal version of the curriculum, along with developing marketing materials. This streamlined workflow ensures that the course curriculum aligns with current industry needs while efficiently integrating human oversight when appropriate.

Customizing Content to Prior Knowledge, Learning Styles, and Preferences

What
Ideally, bootcamp content should be differentiated to provide the best possible learning experience for each participant. This approach requires significant time, effort, and investment. Unfortunately, this crucial aspect is often overlooked. Bootcamp participants often come from diverse backgrounds, each with varying levels of prior knowledge and unique learning preferences. Producing high-quality instructional materials that are both engaging and accessible to all students requires careful consideration of different content types, including visual aids, hands-on activities, interactive elements, videos, and a healthy dose of GIFs. Additionally, the language used should be adapted to each participant not only in terms of complexity but also to accommodate non-English speakers, ensuring clarity and understanding for all.

Impact
Generative AI presents a promising opportunity to move away from the one-size-fits-all approach commonly seen in today's (higher) education, allowing for more personalized learning experiences. Customizing content to align with participants' prior knowledge and diverse learning styles will be transformative for the future of education. Together with an initial assessment, an AI agent can effectively evaluate participants' existing knowledge, strengths, and learning preferences. This approach

will build a unique profile for each student, which can then serve as the foundation for automated content development.

A generative AI technique called retrieval-augmented generation (RAG) can be used to make sure that the student profile is considered while generating content. This technique can be extremely beneficial for creating targeted preparatory work that addresses discrepancies between the program's entry requirements and a participant's prior knowledge, as this will be unique for every participant.

Dynamic Content

What

In a workflow without generative AI, updating content to reflect new developments or incorporating feedback of participants is a labor-intensive process. Ensuring that content is up to date is less straightforward than it seems. Different newsletters need to be read daily, academic papers might need to be checked regularly, or these days keeping a close eye on the X platform might help you to stay up to date on certain topics. This manual approach not only consumes valuable time but also impacts the profit margins of training providers due to the challenges previously discussed. Additionally, if the person responsible for the update differs from the original content creator, inconsistencies in style may arise, which can detract from the learning experience for participants.

Impact

Generative AI excels at recognizing the style of a given text and seamlessly integrating new content that matches that style. Moreover, when a workflow is properly configured, it is able to autonomously scan for necessary updates at regular intervals. By incorporating a human review process, the risk of incorrect updates is minimized. If the AI agent identifies a necessary update, it should present the proposed change to a human reviewer, accompanied by a selection of supporting and credible references. The reviewer will then make the final decision. This approach ensures that the curriculum remains relevant, error-free, and up to date while minimizing the need for extensive manual revisions and constant monitoring.

Content Delivery

What
Learning is inherently non-linear and should be treated as such. However, human teachers often face challenges in accommodating this reality due to the need to adhere to schedules and contractual obligations. Relying on a teacher who lectures for extended periods under time pressure and who delivers a one-size-fits-all session is no longer effective, especially in an era of shortening attention spans and increasing demands for personalized learning.

Impact
GenAI can effectively address this challenge by adapting the instruction materials, the teaching methods, and language based on the participants, ensuring a personalized educational experience. The same logic that was discussed for the creation of tailored preparatory work applies to the content delivery of the bootcamp. Every participant will have the same learning objectives, but the path towards it will be different. For example, some students might be offered video content in Spanish while others might be offered interactive Q&As in English. With generative AI capable of producing content in seconds, the challenge of time-consuming material and content creation is no longer an issue.

Moreover, nearly continuous assessments could be implemented, taking various forms such as quizzes, project prompts, and curriculum-based rubrics. Since these assessments will be generated with GenAI, they can be created almost instantly while ensuring alignment with learning objectives. RAG can be used to make sure that learning objectives are considered while creating assessments. Furthermore, as GenAI can also take on evaluation tasks, feedback can be provided instantaneously and objectively, which is a significant advantage as grading is often considered time-consuming, stressful, and susceptible to bias and errors.

As a matter of fact, a separate AI agent can be responsible for evaluating a candidate's performance in a consistent and coherent manner. This agent can keep track and interpret the progress that a participant is making throughout the program while making sure that answers provided with GenAI by the participant are detected. For deeper understanding of a participant's progress, a GenAI-led interview could be an interesting option. The agent responsible for evaluation could delegate this task to an interview agent. A generative AI-driven interviewing approach has the potential to mitigate human bias, as personal details about participants would not be shared with the chatbot. Furthermore, it offers a more cost-effective and scalable alternative to traditional human interviewing. Given the complexity of this interaction, it may be advantageous to establish a subsystem of AI agents, including a quality assurance (QA) agent that verifies the relevance and clarity of proposed questions. If a question does not meet the necessary standards or doesn't fit with the chat history, it should be rejected.

As a result, the need for human instructors during the content delivery phase should be reduced to rare occasions. While AI can handle much of the instructional load, human involvement remains valuable for providing emotional support, enhancing critical thinking, sharing real-world experiences and ethical discussions.

Conclusion

The integration of generative AI and multi-agent systems presents a transformative future for education, especially in addressing the short-comings of traditional university programs. GenAI allows for more affordable alternatives or additions to traditional university programs. Bootcamps, combined with generative AI, offer a more agile, cost-effective, and skills-focused learning experience that is better aligned with the demands of the modern job market. The automation of content creation, curriculum development, and personalized learning paths through AI significantly reduces the time and effort required to develop high-quality instructional materials. This allows for greater inclusivity, accommodating learners' diverse backgrounds and skill levels. Moreover, the ability to constantly update and customize content ensures that learners receive the most current and relevant education.

The potential of generative AI in content delivery and assessment is equally transformative. By adapting to individual learning styles and providing immediate feedback, AI can create a more dynamic and engaging educational experience. While AI can automate many teacher tasks, the role of human instructors remains useful in fostering critical thinking, bringing real-world experiences and ethical discussions into the learning environment, and offering emotional support. Ultimately, the integration of AI in education not only enhances the efficiency of content delivery but also opens the door to a substantially more scalable model of learning that could reshape the future of education.

As a matter of fact, the massive open online course provider Coursera has started implementing some of the previously discussed technologies. Its CEO Jeff Maggioncalda, along with the executive team, showcased several innovative generative AI products, including:

- *Coursera Coach*: a chatbot designed to understand the context of a learner's journey and provide answers to their inquiries, while ensuring not to reveal specific quiz answers.

- *Course Builder*: a tool utilized by businesses to swiftly customize extensive courses or specializations by selecting the most pertinent sections for their needs.

- *Coach for Interactive Instruction*: this feature enables learners to engage in Socratic dialogue, allowing them to learn or practice new concepts through conversation.

Given the extensive impact of generative AI on education and society, it can be argued that teaching generative AI as a standalone subject to a diverse group of students and instructors could be highly beneficial to society. Prioritizing the democratization of this technology is essential to empower students in their learning experiences and to prevent a handful of powerful companies from monopolizing the landscape.

In this chapter, I focused on using GenAI for higher, tertiary education bootcamps. Considering that individuals in the West spend an average of 15,210 hours in school before turning 18, there is significant untapped potential for implementing Generative AI in the content delivery phase in lower education as well. This is particularly relevant given the widespread shortage of teachers in primary and secondary education. With the substantial time, effort, and financial resources invested in education, huge efficiency gains can and will be reaped in the near future.

About the Author

Cedric De Schaut is a multilingual GenAI and data expert, proficient in data analytics, data science, and data engineering. He has been teaching these subjects through over 50 bootcamps around the world. Having established several bootcamps on what generative AI can do for multinational corporations, he has conceived and taught concrete use cases across different domains, no code and with code. Cedric's bootcamp on Generative AI for Data Scientists is particularly well received, with a net promoter score of 98%. Cedric also has a newsletter. You can find more information on his website.

Email: info@cedric-genai.com
Website: https://sites.google.com/view/cedric-de-schaut-generative-ai/home

CHAPTER 29

ORGANIZATIONAL AI-DRIVEN ENTERPRISES FOR EXCEPTIONAL PRODUCTION AND VALUE

By Stanislav Sorokin
Founder of Bles Software; AI Web Applications
Tel Aviv, Israel

Dedicated to Keren, my best friend, soulmate, and cherished wife who is always by my side, and to Gaia and Noya, our two exceptional daughters. Always very proud of you, always measure the gain—keep learning, growing, training, and reaching new heights every day together!

The impediment to action advances action.
What stands in the way becomes the way.
—Marcus Aurelius

In an era where technology evolves at an unprecedented pace, we stand on the precipice of a revolution that promises to redefine the very fabric of society, business, and entrepreneurship. Artificial intelligence (AI) is a current concept ready to be embraced, having permeated our daily lives and industries, unlocking opportunities previously unimaginable. The transformative power of AI has made it an essential tool for individuals and organizations alike, empowering us to achieve more than ever before.

Imagine a world where a solo entrepreneur can command a powerful network of AI specialists—each possessing the expertise of a Nobel Prize laureate in their respective domains—working tirelessly around the clock. This is not a futuristic dream but a tangible reality unfolding before us right now. The integration of AI into organizational structures heralds a new paradigm, one where businesses can scale exponentially while maintaining a lean human workforce. This is the era of boundless opportunities where AI serves as the ultimate catalyst for unprecedented growth and success.

The Exponential Leap Forward

The traditional constraints of scaling a business—limited human resources, time, and capital—are being obliterated by the advent of AI. Companies can now deploy thousands, even millions, of AI agents capable of performing complex tasks with unparalleled efficiency. These AI employees operate at productivity levels of 20+ times greater than their human counterparts, executing tasks ranging from software development to advanced research. With AI as an integral part of your operations, achieving levels of productivity that were once thought impossible is now within reach.

Consider a medium-sized enterprise with 70 human employees. By integrating AI, this company can augment its workforce with a legion of AI agents, each specialized in different domains. The result is a hyper-efficient organization that can operate 24/7, breaking barriers of intelligence, time zones, and human limitations. The power of AI is not only in its ability to perform but also in its relentless nature—it works tirelessly to propel your business forward, ensuring that no opportunity is missed.

Imagine that instead of traditional limitations, your company is empowered to do more with less. AI takes care of the repetitive and mundane while your human workforce focuses on creativity, strategy, and innovation. With AI, your company becomes capable of leveraging data at an unprecedented scale, making well-informed decisions that fuel business growth. The key lies in this unique collaboration between human intuition and AI precision—creating a synergy that drives unparalleled results.

Reimagining Organizational Structures

The infusion of AI into businesses necessitates a rethinking of organizational hierarchies. Human employees transition into strategic roles—visionaries who steer the company's direction, uphold ethical standards, and foster innovation. They have access to a central management interface—a comprehensive dashboard that provides real-time insights into operations, performance metrics, and AI employee activities.

Communication becomes seamless through a simple chat interface accessible via smartphones. Executives and managers can interact with AI coordinators, receive updates, and issue directives on the go. This system allows for agile decision-making, with AI agents autonomously handling day-to-day operations based on predefined parameters and live data analysis. The newfound agility enables your business to respond to changes in real time, adapt strategies as needed, and continuously move toward achieving your vision.

AI also transforms how you visualize your business operations. Imagine having a clear, real-time overview of every aspect of your enterprise, with AI providing insights and analysis that empower you to make decisions confidently. No longer confined by traditional boundaries, your company evolves into a dynamic, data-driven powerhouse, capable of expanding into new markets and pushing beyond established norms.

A Vivid Illustration: Your AI-Enhanced Company

Let's explore how you could transform your business vision into reality by harnessing the power of AI.

The Genesis

Imagine you decide to launch your own tech venture. You recognize an exciting opportunity: by combining your innovative vision with AI capabilities, you can create something extraordinary in the software industry. You see how AI could amplify your team's talents and unlock unprecedented scalability. You decide to dive headfirst into this endeavor, knowing that the potential rewards are limitless.

The idea is simple yet profound: leverage AI to transform your operations, empower your workforce, and innovate without limits. You recognize that embracing AI is not merely a choice; it is the gateway to unparalleled possibilities. With AI by your side, you step into the future of business, equipped to make a powerful impact.

The Integration
You strategically weave AI agents throughout your operations:

- *Software Development*: your human programmers collaborate with AI developers, creating a dynamic force that transforms coding, testing, and deployment. This partnership turns ambitious development timelines from months into days, empowering your team to achieve more than ever before. The AI operates as an extension of your developers, suggesting improvements, identifying bugs, and speeding up every aspect of the software development cycle. Looking ahead, AI will evolve to fully automate all coding tasks, with AI-driven managers overseeing the development cycle. These AI managers will coordinate and communicate seamlessly with other AI teams across your organization, creating a unified and frictionless development environment.

- *Market Research*: your AI analysts become tireless market explorers, discovering global trends, customer insights, and emerging opportunities. This rich, real-time intelligence empowers you to shape market-leading strategies with confidence. By using AI to sift through data at unimaginable speeds, you position yourself at the forefront of your industry, consistently leading rather than following trends.

- *Customer Service*: you implement AI chatbots that complement your support team, offering around-the-clock assistance with a personal touch. Advanced natural language processing ensures every customer interaction reflects your commitment to excellence. As a result, customers feel heard, valued, and understood. The satisfaction that comes from this level of service translates to brand loyalty and a competitive edge.

- *Financial Management*: AI financial agents using tools become your strategic advantage, spotting patterns in transactions, identifying optimal budget allocations, and uncovering promising investment possibilities. With these agents handling every aspect of financial data analysis, you and your employees have a dedicated dashboard to review results in real time and fine-tune your organizational AI workforce for maximum efficiency.

AI Chief Management and Complete Automation

At the top of your organization's AI ecosystem stands a powerful reasoning AI Chief Manager, orchestrating every process across departments and ensuring total alignment with your strategic objectives. Through full-scale automation, this AI Chief oversees a network of specialized AI agents—each managed by its own AI coordinator—who collectively streamline workflows, eliminate inefficiencies, and drive growth at an unprecedented scale. Meanwhile, you and your employees gain real-time visibility into all AI teams, their progress, analytics, and actionable insights through a dedicated dashboard. This unified view empowers everyone to monitor performance, make data-driven decisions, and swiftly adapt strategies across the entire organization. By integrating AI Chief Management with comprehensive dashboards and automated processes, you transform your operations into a seamlessly coordinated machine—one that consistently delivers optimal results with minimal oversight.

The Outcome

Through this powerful AI-human collaboration, your company soars to new heights. You expand globally without traditional constraints, as your AI systems masterfully adapt to diverse markets and regulations. Your human team focuses their creativity on groundbreaking innovations and strategic relationships, driving your company toward ever-greater achievements.

AI empowers your enterprise to operate seamlessly, ensuring that no opportunity is overlooked. Imagine being able to identify a market trend before your competitors or launching a product in half the time it would have taken previously. AI enables these feats by delivering data, insight, and action at lightning speed. This means your company is no longer reactive—it becomes proactive, shaping the industry and defining its trajectory.

With AI as your partner, the future becomes an exciting landscape of boundless opportunities. Scaling your business to international markets no longer requires massive infrastructure or large investments. Instead, your AI systems enable you to enter new territories, understand market dynamics, and tailor your approach to meet diverse consumer needs, all while optimizing your resource allocation.

Starting Your AI Transformation Today

Embarking on this journey doesn't require an immediate overhaul of your existing business. You can begin by identifying areas within your operations that are ripe for AI integration. Assess your business processes to pinpoint tasks that are repetitive, data-intensive, or could benefit from automation. Common areas include customer service, data analysis, marketing, and supply chain management.

Investing in AI infrastructure is crucial for this transformation. Building an AI-ready environment involves acquiring the necessary hardware, software, and networking capabilities. Cloud-based solutions offer scalability and flexibility, reducing the need for significant upfront investment. The path forward begins with incremental changes, each one empowering your business to grow stronger and more capable.

It's also essential to upskill your workforce. Educate your team about AI technologies by providing training programs that help them understand how to work alongside AI agents effectively. This not only improves collaboration but also alleviates fears about job displacement. Empowerment comes from knowledge—by equipping your team with the skills to utilize AI, you ensure they remain integral to your company's success.

Partnering with AI specialists can greatly enhance your integration efforts. Collaborate with experts who specialize in AI; they can provide valuable insights, customize solutions to your needs, and ensure a smooth transition. These partnerships open doors to advanced capabilities that drive growth and bring your vision to life.

Implementing AI should be done gradually. Start with pilot projects to test applications within your business. Monitor performance, gather feedback, and refine your approach before scaling up. With each successful implementation, your confidence in AI grows, and so does the potential to revolutionize your operations.

The Ethical Imperative

With the immense power that AI brings, businesses have a moral obligation to ensure that their deployment of AI technologies is ethical and beneficial to all stakeholders and society. AI systems should be designed to eliminate biases, promoting fairness in decision-making processes. This involves using diverse datasets and regularly auditing AI outputs for unintended discrimination.

Maintaining transparency about how AI agents operate within your organization is crucial. Clear communication builds trust with customers, employees, and partners. Safeguarding sensitive data through robust cybersecurity measures is equally important. Respecting privacy laws and regulations is not just a legal requirement but a commitment to your customers' trust.

Establishing protocols for accountability when AI agents make errors is essential. Having human oversight ensures that there are mechanisms to correct mistakes and learn from them. Moreover, businesses should leverage AI to make positive impacts beyond their operations. This could involve using AI for environmental sustainability, supporting social causes, or improving accessibility for underserved communities. By using AI to contribute to the greater good, businesses create value that extends beyond profit—fostering a future that benefits all.

A Collaborative Future

The integration of AI doesn't signify the replacement of humans but rather an evolution toward a collaborative ecosystem where humans and AI complement each other's strengths. AI provides data-driven insights that enhance human judgment. Together, they can make more informed decisions that propel the business forward.

With AI handling routine tasks, human employees can focus on creative problem-solving, innovation, and strategic planning. AI agents can also operate across different time zones and languages, allowing businesses to expand their global footprint without significant additional resources. This new synergy empowers individuals, allowing them to utilize their unique strengths in ways that were previously not possible.

Imagine the creativity unlocked when your team no longer needs to spend hours on tedious tasks. The ideas that flourish, the innovations that emerge, and the solutions that are created will set your company apart.

AI is the ultimate enabler, freeing up human potential and allowing your workforce to achieve things beyond their current imagination.

Preparing for the Next Frontier

As AI continues to evolve, staying ahead requires continuous learning and adaptation. Keeping abreast of the latest AI developments and trends will help you anticipate changes and adapt your strategies accordingly. Encouraging experimentation and creativity within your organization fosters a culture that embraces change, making it more likely to thrive in a technologically advanced landscape.

Engaging with the AI community by participating in industry forums, conferences, and networks can spark new ideas and collaborations. This interaction not only broadens your knowledge but also positions your business within the forefront of AI innovation.

By fostering a culture of curiosity and embracing the transformative potential of AI, your business is not just preparing for the next frontier—it is defining it. With AI as your ally, the only limit is the scope of your imagination. Together, we can build a future that is intelligent, compassionate, and full of endless opportunities.

Conclusion: Embracing the Responsibility

The potential to build AI-driven organizations that can scale infinitely is both exciting and daunting. It offers the promise of unprecedented growth, efficiency, and innovation. And it also places a significant responsibility on business leaders to wield this power ethically and thoughtfully.

Our decisions today will shape the world of tomorrow. By embracing AI with a commitment to positive impact, we can build a future where technology enhances the human experience, drives progress, and creates opportunities for all. Because with great power comes great responsibility.

About the Author

Stanislav "Stas" Sorokin is an entrepreneur, visionary, and expert in software development. As the founder of Bles Software, Stas leads a team dedicated to creating innovative SaaS solutions and applications that make a global impact. With over a decade of experience in the tech industry, he has built a reputation for delivering cutting-edge, client-focused solutions that leverage AI and advanced technologies. Bles Software manages more than 55 projects simultaneously across platforms like Fiverr and Upwork, where Stas is recognized as a top seller and trusted professional.

Stas's entrepreneurial philosophy is rooted in his belief in creating "technological blessings"—rare, transformative innovations inspired by the concept of the black swan. His leadership in developing scalable software and AI-driven systems positions Bles Software as a key player in driving the next wave of technological revolution.

Beyond his professional achievements, Stas is passionate about personal development and health. He is an advocate for intermittent fasting, longevity-focused supplementation, and disciplined fitness routines. As a dedicated parent, he encourages his children to explore their passions and excel in their pursuits. Stas is also a mindset enthusiast, implementing frameworks like "Extreme Ownership" and B.J. Fogg's behavior models (You can see more in "Tiny Habbits" book) to enhance personal growth and productivity. He is a lifelong learner, committed to improving himself and inspiring others to achieve their fullest potential.

In addition to his work, Stas has authored thought-provoking content, including "Embracing Moore's Law Squared: How to Build Universally Expansive Businesses in the Age of Exponential AI," where he shares insights on scaling businesses in the rapidly evolving AI landscape.

Bles Software

Bles software is a pioneer in the realm of artificial intelligence integration, specializing in embedding advanced AI solutions into business organizational systems. By implementing agentic frameworks that facilitate autonomous decision-making, Bles Software transforms traditional business models into intelligent, adaptive entities capable of thriving in the modern digital landscape. Their expertise in AI not only streamlines operations but also empowers businesses to make data-driven decisions with unprecedented speed and accuracy. With a commitment to innovation and excellence, Bles Software equips organizations with the tools they need to lead in an era of exponential AI growth. Discover how Bles Software can elevate your business by visiting bles-software.com or send an email to info@bles-software.com.

Email: stas4000@gmail.com

LinkedIn: https://www.linkedin.com/in/stas-sorokin
For opportunities and supercherning your business with organizational
AI contact Stas directly on WhatsApp at +972509077339.

CHAPTER 30

AI TAUGHT ME TO GARDEN: PRACTICAL APPLICATIONS FOR LEARNING ANYTHING

By Matthew Switzer
AI Solutions Strategist, UX Researcher
Nederland, Colorado

> *"If you have a garden and a library,*
> *you have everything you need."*
> —Cicero

Spring on the Horizon

March snow fell through the pines, blanketing the deep green silhouette of Mount Thorodin several miles to my south. I stood by the window of my home, nestled in Colorado's Front Range, taking in the last moments of winter. The snow would soon melt, and the ground soften. I had never planted a vegetable, but I had already decided: this would be the year I grew a garden.

I knew gardening at 8,200 feet, with a short four-month growing season, would bring its share of challenges. June snow storms were entirely possible, and hungry moose, elk, and rabbits were almost certain. My goal—one I hadn't fully grasped—was to grow my own groceries, hoping to offset rising food costs. I had the land, the gumption, and just the right balance of naivety and determination to believe I could make it happen.

In my true nature, I'm a dreamer, often setting ambitious goals. I envisioned a flourishing garden—rows of lush lettuce, ruby-red tomatoes, and kitchen counters overflowing with freshly harvested produce. Yet, I had little idea how to bring that vision to life. Still, I've learned that even the loftiest goals can be reached by breaking them down into small, achievable steps.

As a researcher, I approached the project by identifying what I didn't know. I Googled phrases like "high-altitude gardening" and "cold-tolerant produce," but the results seemed unreliable. Near-freezing mornings and high-UV afternoons didn't match the advice I found. I was determined to get my garden right the first time and, respectfully, couldn't afford to rely on the YouTube advice of a garden enthusiast from the New York Catskills.

I soon realized that the information I needed simply didn't exist in one place. I had gathered bits and pieces of useful information, but none was considered holistically with my unique environment in mind. That's when an unlikely idea occurred to me: could the generative AI I used for research be applied to high-altitude agriculture? Could AI actually teach me how to grow a garden?

Cultivating Knowledge

When it comes to generative AI, my go-to tool is OpenAI's ChatGPT. GenAI creates new content by analyzing patterns from vast amounts of data, acting as a powerful platform for co-creation and problem-solving. ChatGPT's versatility extends far beyond tasks like drafting emails or polishing resumes; it can tackle complex challenges and provide practical solutions, with its potential limited only by our creativity. I saw an opportunity to use it to better understand and manage the unique conditions of my garden—and decided to make the most of it.

Shortly before I began my research, ChatGPT introduced a feature allowing users to teach it in a way I liken to educating an intern. You provide materials, define reliable sources, and give instructions on how to interpret the data. The AI learns, refines, and evolves based on the context and information it receives. The more I thought about it, the more it seemed possible that GenAI could help me not just theoretically, but practically.

I sat at my desk, a dozen browser tabs open, reflecting on what I knew about generative pre-trained transformers, more commonly known as GPTs. All I had to do was feed my model the best resources to help it teach me to garden and ask the right follow-up questions. I needed to build a focused ecosystem of information to guide my journey. I've found that the best AI results come from teaching it how to teach us. Feeling optimistic, I opened the GPT editor and typed simple instructions, setting parameters based on what I already knew:

> *"Help me understand and plan for a vegetable garden in high altitude Colorado. I live at 8,200 feet elevation and am in growing zones 4b–6a. Only reference sources that explicitly reference Colorado, high altitude farming and gardening, or my growing zones. Cite your sources."*

Then, I asked myself: "What knowledge would an experienced gardener who can answer all my questions have?" With that thought, I began gathering resources—PDFs from Colorado State Universities on high-altitude gardening practices, climate data, and my notes from chatting with more seasoned gardeners in my neighborhood. I fed these materials into my GPT's "knowledge" section, ensuring it had access to the most relevant, location-specific expertise to guide me.

With my gardening GPT primed, I asked it the most fundamental question: "What plants should I grow?" The model responded with detailed advice:

> *"Given your high-altitude location at 8,200 feet in Colorado (growing zones 4b–6a), your garden will face a short growing season, cool nights, intense sunlight, and potentially late and early frosts. You'll want to focus on cold-hardy, fast-maturing plants that can thrive in such conditions. Additionally, choosing crops that are drought-tolerant and resilient against temperature fluctuations will be key."*

It then proceeded to list ten vegetables suited to my conditions. I double-checked everything—it was spot on. The varieties had me brimming with excitement—Emerald Fan Lettuce, Purple Dragon Carrot, Black Plum Tomato. I was filled with visions of a vibrant rainbow garden. With a sly grin, I asked where I could buy organic seeds for these varieties. Within moments, I was browsing seed supply sites, filling my shopping cart.

I updated my GPT's instructions with the specific vegetable varieties I selected, along with their growing instructions. The more specific I was with the information I provided, the better the results. It was like assembling a puzzle—starting with the essential edge pieces, adding layers of detail, and helping the AI connect each piece to form a complete picture. In the process, I felt like I wasn't just teaching the AI; it was also guiding me on how to approach and care for my garden.

As I interacted with the GPT, I began to realize it was about more than simply inputting disparate data and hoping for answers; it was about building a dialogue with the AI, refining and adjusting the information until I felt confident in the advice. AI can be an expansive creator, pulling indiscriminately from a vast pool of data; or, with our guidance, it can be precise and focused, delivering exactly what's needed.

Planting the Seeds of Success

Planting my seeds felt like the true beginning. The first seed I planted was Sputnik Arugula. According to my GPT, it was a fast grower, cold-tolerant, and the first plant ready for transplanting outdoors. I balanced the tiny brown seed on my pinkie, feeling the weight of its potential, and gently pressed it into the soft, dark soil. I placed the seed tray under a modest grow light in the corner of my office. Outside, the cold nights still brought frost, but indoors, I had created a tiny pocket of warmth and fertility.

Some ten days later, I stepped into my office and shouted with joy when I was greeted with green sprouts triumphing through the soil. Seeing the fragile seedlings reach for the light sparked a profound sense of wonder within me. Yet, the reality of gardening at 8,200 feet soon hit. The growing season is short—a tight four-month window racing against time for each plant to mature. My seedlings needed a head start.

My single grow light quickly became insufficient. Within a couple of weeks, I had over 100 seedlings spanning 18 plant varieties. The shelf in my office just wasn't cutting it anymore. ChatGPT recommended expanding my setup, and soon I found myself clearing out a spare room to install a full walk-in grow room—a five-by-five-foot space bathed in LED light, with fans circulating air to nurture the seedlings before they faced the intensity of high-mountain weather.

ChatGPT, the Gardener

As the weather warmed and the soil softened, I began digging. Cool earth stained my knees and sifted through my fingers as I shaped the garden beds. The space felt alive with possibility. I opened the ChatGPT app on my phone, enabling voice interaction, and began speaking to it as I would a trusted mentor. With its help, I knew exactly how far apart to space my lettuce, which plants thrived together, and which needed distance. It recommended planting cilantro near my peppers to repel pests with its strong scent while pairing beans with squash to naturally enrich the soil with nitrogen. Every decision was rooted in solid research and collaboration.

I followed its guidance down to the inch, marveling at the partnership that had developed. The AI wasn't just spitting out random facts—it had become an integral part of my process. My garden grew greener and fuller with each passing day, promising the success I had envisioned months ago.

Not everything went according to plan, though. It was my row of leafy greens that taught me my first real lesson in failure. I needed a strategy to move my garden outside before morning temperatures rose above freezing, so I asked, "How can I protect my varieties of leafy greens from frost?" It responded,

> *"Leafy greens like lettuce, spinach, kale, and chard are somewhat frost-tolerant but will benefit from extra protection to extend their growing season and prevent damage ..."*

It then suggested several methods, first being to cover my rows with a hoop house—a small tunnel-like greenhouse that would shield plants from the still-chilly mornings. However, ChatGPT, or I, hadn't accounted for the early June heat wave. The afternoon sun trapped beneath the plastic turned my tender lettuce into shriveled, wilted leaves. It was a rookie mistake, and I stood in the garden, staring at the limp plants, feeling the sting of a lesson learned too late. The AI had advised on planting schedules and spacing, but I hadn't considered telling it about local microclimate oddities like sudden heat waves. It was a gap in the system—one I hadn't anticipated because I hadn't provided that critical context.

Failure, though, is just part of learning, for both me and the AI. I took that mistake and fed it back into the system. I updated my GPT with current data from local weather patterns and instructed it to account for environmental extremes. It wasn't just about teaching myself to garden— it was about teaching the AI from my own failures, refining the model so that it could warn me about heat waves or suggest alternative strategies for fragile plants. I learned that growing food isn't just about under- standing what to do; it's about adapting quickly when nature throws the unexpected your way. My lettuce didn't survive, but that misstep ensured the next crops would stand a better chance.

As the summer unfolded, my garden became a living classroom, with ChatGPT as my ever-present professor. Each morning, I'd walk outside, phone in hand, and observe the subtle changes in the plants—new leaves unfolding, tiny buds forming, or the occasional pest munching away at something they shouldn't. I'd talk through each observation with my GPT, troubleshooting like two partners solving a complex puzzle. When the soil dried out faster than anticipated during a hot spell, the AI recom- mended mulching to retain moisture, a solution that quickly improved the health of my struggling crops. When a mid-July hailstorm flattened half of my tomato plants, ChatGPT guided me in pruning the damaged branches and suggested protective measures to help the survivors recover. It was a constant cycle of learning and adapting—just as I taught the AI with updates about my microclimate, it taught me how to pivot when nature didn't follow the textbook.

The process wasn't without setbacks. Some plants didn't thrive as expected, and I encountered several surprises that the AI hadn't accounted for, like an unexpected invasion of chipmunks that nearly decimated my bean patch. But each failure offered insight, one I shared with ChatGPT by adjusting its instructions or feeding it new information. It felt like an ongoing dialogue, a back-and-forth exchange of data and solutions. Together, we navigated the complexities of high-altitude gardening. By the end of the summer, I had deepened the GPT's knowledge, shaping it to better navigate my garden's unique challenges. In turn, I, too, had grown—no longer just a novice, but a gardener in tune with the rhythms and personality of my land.

As autumn crept in and the aspens shimmered gold, I stood in my garden—now a patchwork of success stories—and reflected on the months of dedication that had led to the fruits of our labor—ChatGPT and me. Rows of squash, ripe tomatoes, and waste-high potato stalks stood resilient against the season's end, their growth a testament to months of collaboration between me and artificial intelligence. I had done it. I had successfully grown a garden at 8,200 feet and did so using only the guidance of AI. This garden wasn't just a triumph of nature, but a demonstration of how technology, when used thoughtfully, can open doors to skills previously out of reach. I had learned to grow a garden—perhaps a little differently than most—by tailoring a knowledge resource to my needs and drawing from it as my continuous mentor.

It was humbling to realize that I hadn't needed years of apprenticeship or a stack of gardening books to get here. Instead, I leaned on a tool that could synthesize expertise, cross-reference research, and adapt to new challenges, all while evolving alongside me. This was more than a digital assistant—it had become a co-creator in my success, learning from my mistakes and refining its advice with each setback overcome. If someone had told me two years ago that I could grow a bountiful garden using AI as my sole guide, I would have had more questions than belief. Now, with my tomato plants at shoulder height, I know it to be true.

AI for Good, at Scale

Reflecting on my experience, one thought kept resurfacing: what I had achieved on a small scale could be applied on a broader level with a more significant impact. I knew this methodology could seamlessly extend to the corporate world; I was already using generative AI regularly to cross-reference mixed-methods research, creating specialized GPTs tailored to each project. Just as I had curated high-altitude gardening advice to suit my unique needs, workplace teams can refine AI models to address complex challenges—whether in project management, research synthesis, or strategy development. It's not about AI replacing human expertise but augmenting it, enabling us to make informed decisions more quickly and with greater precision.

More importantly, the curiosity lingered that if ChatGPT could help me, a novice, navigate the complexities of agriculture, what could this technology do for entire communities grappling with environmental and agricultural challenges? Just as I had customized my AI to guide me through unpredictable growing seasons, farmers in regions with extreme climates or limited resources could do the same. By feeding local data into AI models—such as weather patterns, soil conditions, and indigenous practices—they could access practical, location-specific advice to improve crop yields and manage resources more effectively. The possibilities seemed limitless, from applications in urban agriculture to combat local food insecurity to addressing broader humanitarian crises, like ensuring access to clean water in vulnerable regions.

The theoretical applications of AI stretch far beyond agriculture. Imagine a world where generative AI becomes a cornerstone for addressing global challenges across disciplines. In medicine, for example, AI could be trained to assist healthcare professionals in underserved regions by suggesting treatment plans tailored to limited medical resources. By adapting to geo-specific constraints AI could provide lifesaving solutions where traditional methods fall short. Just as I had trained ChatGPT to navigate my garden's challenges, these systems could be tailored to tackle some of humanity's most pressing problems, offering powerful adaptability at scale.

In education, generative AI could serve as personalized tutors for students worldwide, bridging gaps in access to quality teaching and helping learners in under-resourced areas overcome barriers to education. By tailoring to individual learning styles and needs, AI could break down barriers to education, empowering learners with opportunities previously out of reach. These aren't far-fetched ideas—they are practical and achievable if we apply the same creativity and resourcefulness to other domains as I did with my garden. AI, when designed thoughtfully, isn't just a tool for solving problems; it's a tool for empowerment, amplifying human potential where resources are scarce.

As I look back on my journey, one truth feels undeniable: the partnership between humans and AI has the power to reshape how we approach the unknown. Whether it's growing squash at 8,200 feet or finding innovative ways to mitigate global challenges, the key lies in collaboration. AI can amplify our abilities, enabling us to push boundaries we never thought possible. My garden stands as proof of what can be accomplished when technology is guided with intention and curiosity. If one person, in one corner of the world, can achieve this with AI, imagine the collective potential waiting to be unlocked. The future isn't about AI doing things for us—it's about what we can achieve together.

About the Author

Matthew Switzer is an AI solutions strategist with a background in UX research and product design, known for his human-centered approach to innovation. Passionate about pushing the boundaries of AI's applications, Matthew develops tailored solutions empowering individuals and organizations to solve complex challenges—ranging from optimizing processes to exploring creative and innovative use cases. His work bridges cutting-edge technology with real-world impact, demonstrating how AI can serve as a tool for empowerment, not just automation. Whether guiding teams in streamlining analysis or helping individuals tackle niche challenges, his mission is to expand AI's transformative potential.

Interested in collaborating or leveraging Matthew's AI expertise? Visit his website or reach out directly to start the conversation.

Email: hello@matthewswitzer.io
Website: www.matthewswitzer.io

CHAPTER 31

AI IN ENGINEERING AND ARCHITECTURE

By Sakina Syed, B.Sc.
AI Engineer and Consultant
Toronto, Ontario, Canada

Artificial intelligence is not a substitute for human intelligence;
it is a tool to amplify human creativity and ingenuity.
—Fei-Fei Li

Artificial intelligence (AI) is not just a technological advancement; it's a transformative force reshaping the landscape of engineering and architecture. Imagine a world where buildings are designed with incomparable precision, where construction developments are completed faster and more carefully with higher standards of safety, and where the creative potential of architects and engineers is set free like never before. With AI, the future of our built environment is not just a vision, but an imminent reality.

AI can design thousands of different engineering solutions in a fraction of the time it would take a human. For example, generative design software uses AI to create numerous design options based on specific parameters, constraints, and goals. This technology has been used to design everything from lightweight airplane parts to innovative architectural structures.

AI is revolutionizing the fields of engineering and architecture by enhancing productivity, precision, and creativity. This is the promise of AI and we will explore how to use these tools to our advantage when it comes to designing buildings, cars, and more.

A Brief History of AI in Engineering

It is often thought that AI is new to the field of design and engineering. However, AI has been utilized in engineering since the 1970s. Finite element analysis (FEA) simulations, which are mathematical models predicting mechanical behaviors like stress, strain, heat transfer, and fluid flow, have been in use since the 1970s, initially run via punched cards. The integration of AI into architecture has progressed since around 2010, with architectural firms beginning to explore AI's potential in design and planning. In that era, AI tools analyzed extensive datasets to optimize building designs, improve energy efficiency, and predict maintenance needs.

Fast forward to today, we are in the era of generative design. Technologies such as virtual twins and machine learning are now employed to create more sustainable and efficient buildings. Generative design is a feature of computer-aided design (CAD) applications that automatically generate multiple design alternatives based on specified constraints. This approach is a game-changer for addressing complex engineering problems, as it operates without the need for direct input from engineers, allowing them to focus on other tasks. Traditional design methods often face limitations due to time constraints, limited resources, or the bounds of human creativity. Generative design in engineering overcomes these hurdles by using powerful computational algorithms to explore a vast array of design possibilities. Engineers can then select which designs to explore further, speeding up the design process without requiring detailed attention from the engineer. Both engineers and architects can leverage generative design to achieve fast and precise drawings that they can further refine, design, and transform.

Introduction to AI in Engineering and Architecture

In the field of engineering, AI has countless applications enabling engineers to stay ahead of the curve and maximize their design potential.

276

Applications range from predictive maintenance and intelligent design to simulation, autonomous systems, and natural language processing (NLP) for engineering documentation. AI also enhances quality control and sustainability assessments, improving every aspect of the engineering process. Whether in mechanical, civil, structural, or other engineering disciplines, AI assists in the design and development process, offering valuable support across the board.

Moreover, machine learning algorithms are being used to optimize various engineering processes. For instance, in materials engineering, AI can analyze vast amounts of data to identify new materials with desired properties, significantly speeding up the discovery process. In civil engineering, AI tools are used for structural health monitoring, where they analyze data from sensors placed on infrastructure to detect and predict potential issues.

For example, AI tools such as Azure IOT Central and IBM Maximo allow for better results for engineers in predictive maintenance, which can prevent costly equipment failures to quality control systems that ensure every component meets exacting standards. Intricate systems and processes are optimized with AI's ability to analyze vast amounts of data and make real-time adjustments, leading to more efficient and reliable outcomes.

Architecture is also experiencing a transformation with the help of AI. AI assists architects in generating innovative designs that push the boundaries of creativity while optimizing space utilization and ensuring sustainability. Imagine a design process where AI algorithms suggest the most efficient layouts, materials, and energy solutions, creating structures that are not only beautiful and have aesthetic plus points but also environmentally friendly. AI algorithms can analyze vast datasets of architectural styles, building codes, and environmental factors to generate multiple design options. AI-powered design generative tools also allow architects to list parameters such as spatial restrictions, materials, and building design requirements and environmental conditions to yield results that will factor all of these into the designing process and end results. We will explore tools that can do this a little later.

The Benefits of Integrating AI in Design

AI meaningfully enhances the efficiency of design processes by automating repetitive tasks like blueprint drafting, 3D modeling, and rendering, allowing engineers and architects to focus on more complex and creative aspects of their projects. By using machine learning and pattern recognition, AI can quickly generate design alternatives, optimize layouts, and predict potential issues, speeding up the design process and reducing human error. Integrating AI into design and construction not only accelerates project completion and reduces costs but also improves safety, enabling professionals to create smarter, more sustainable, and aesthetically pleasing structures that showcase human ingenuity and technological prowess.

GenAI in Engineering and Architecture

AI can integrate a multitude of domains, allowing us to approach any field with the combined creative power and understanding of multiple disciplines. Let's quickly define generative AI in the context of engineering and architecture.

GenAI uses advanced algorithms to create innovative, robust, and eco-friendly designs by learning from existing data. Designers input constraints and goals into CAD software, which uses generative algorithms to produce numerous design alternatives. The benefits of generative design include more efficient material use, reduced production costs, and enhanced product performance. GenAI allows engineers to focus on refining the best designs, accelerating the design process, and pushing the boundaries of product design and engineering.

In architecture, it optimizes space usage while in engineering, it improves parts and systems, reducing costs. Generative design allows engineers to input design goals and constraints and define relationships between design elements, and the AI generates multiple design alternatives, optimizing for factors like weight, strength, and cost. For instance, Airbus, the aerospace corporation, used generative design to create a lighter partition for its A320 aircraft, resulting in significant fuel savings.

AI in Architectural Design

Enhancing Creativity and Efficiency in Architectural Design with GenAI
With AI, architects can push the boundaries of traditional design, creating structures that are both functional and aesthetically groundbreaking while maintaining engineering codes and standards. Let's explore some buildings designed with the help of AI. I encourage you to check out these spectacular structures online to see their magnificence.

Examples of Buildings Designed by AI
Shanghai Tower

This skyscraper in China utilized generative AI for various aspects, including optimizing energy efficiency, wind resistance, and material selection. AI-driven simulations were instrumental in designing its distinctive twisting shape, which effectively minimizes wind loads.

The Edge, Amsterdam

Known as one of the most sustainable office buildings in the world, the Edge uses AI to optimize energy consumption. Its architects at PLP Architecture used AI to design its energy systems. It uses sensors and data analytics to adjust lighting and climate control according to occupancy and external conditions, making the Edge one of the greenest and most efficient buildings globally.

Microsoft's Redmond Campus Modernization Project

Image Designed by Sakina Syed

The project uses AI to create a sustainable and smart campus. AI-driven design tools helped in planning the layout, selecting eco-friendly materials, and integrating energy-efficient solutions. This approach not only results in beautiful and functional buildings but also significantly reduces environmental impact, aligning with the company's commitment to sustainability.

The Role of AI in Design Automation

AI-Powered Tools in Engineering
Several AI-driven design tools exemplify the power of automation in engineering and architecture, from assessing environmental factors to designing buildings previously deemed impossible. An example of a tool that does this is Autodesk AI, which integrates into many tools including AutoCAD, Revit, BIM 360, Fusion 360, Autodesk Construction Cloud, and more. For instance, Autodesk AI is a design tool that optimizes design layouts. Construction IQ, part of Autodesk AI, predicts, prevents, and manages construction risks. It does so by analyzing complex project data to offer insights related to quality, safety, cost, and schedule. AI algorithms can help in generating standard details, recognizing objects, and automating annotations.

Digital twins, virtual replicas of physical systems, work with AI to enhance engineering processes by providing real-time simulations and predictive analytics. AI processes data from sensors embedded in physical objects to update the digital twin, allowing for accurate modeling and scenario testing. For example, the city of Singapore has developed a digital twin to simulate urban planning scenarios, improving infrastructure management and sustainability. This combination helps engineers optimize designs, predict maintenance needs, and improve operational efficiency. By continuously learning from real-world data, AI-enhanced digital twins enable more informed decision-making and innovation. Here are some of the tools mentioned above, along with more notable AI design tools that are transforming design automation in architecture and engineering that you can further explore:

1. **Autodesk Generative Design**
    This tool uses AI to explore a wide range of design possibilities based on specified constraints and goals, such as material efficiency and structural integrity.

2. Spacemaker

An AI-powered tool that helps architects and urban planners optimize building designs by analyzing environmental factors and spatial configurations.

3. ARCHITEChTURES

This tool leverages AI to streamline residential planning by analyzing site conditions, climate dynamics, budget constraints, and client aspirations to generate optimal design options.

4. Veras

Developed by Evolve Labs, Veras integrates with popular architectural software like SketchUp and Revit to transform 3D models into photo-realistic renderings using AI.

5. Maket.ai

A generative design software that automates the creation of residential plans, quickly generating custom floor plans and 3D renders that consider zoning codes and client requirements.

6. Midjourney

An AI-powered text-to-image generator that converts text prompts into photorealistic images, aiding architects in visualizing and refining design concepts.

7. Dreamhouse AI

Designed for interior design, this tool offers quick solutions for room transformations and virtual staging by utilizing uploaded photos.

8. Grasshopper for Rhino

A visual programming language integrated with Rhino, which allows for generative design and parametric modeling.

9. *Houdini*

Known for its use in visual effects, Houdini also offers powerful generative design capabilities for architectural modeling.

10. Copilot

Copilot enhances architecture and engineering by generating design ideas, optimizing structural analysis, automating documentation, and facilitating collaboration through intelligent insights and recommendations.

SAKINA SYED, B.SC.

Case Studies of AI Applications in Various Engineering Fields

In the field of mechanical engineering, AI has been great at transforming traditional practices. A notable example is General Electric's use of AI for predictive maintenance in their jet engines. By analyzing data from sensors, AI algorithms can predict potential failures and maintenance needs, reducing downtime and improving safety. This application has not only enhanced the reliability of their engines but also resulted in significant cost savings. Another example is Siemens, which uses AI-driven simulation software to optimize the design and performance of their products. This software allows engineers to test various scenarios and make data-driven decisions, leading to more efficient and innovative designs.

In civil engineering, AI is being used to monitor and maintain infrastructure. The city of Los Angeles, for instance, has implemented an AI-based structural health monitoring system for its bridges. This system uses sensors to collect data on vibrations and movements, which AI algorithms analyze to detect any signs of structural weakness or damage. This proactive approach helps in preventing catastrophic failures and ensures the safety of the infrastructure. In the realm of environmental engineering, AI is being used to optimize water treatment processes with NLP or natural language processing. Companies such as Xylem are using AI to analyze data from water treatment plants to improve efficiency and reduce energy consumption.

AI Robots in Engineering

AI-driven robotics enhances construction efficiency and safety. Boston Dynamics' Spot robot, for instance, is used on construction sites to perform routine inspections and capture data, reducing the risk to human workers and speeding up project timelines.

Robotics and automation, powered by AI, are also playing a crucial role in manufacturing and assembly lines, improving precision and reducing human error. These AI tools collectively contribute to making engineering processes more efficient, reliable, and innovative, paving the way for advancements that were previously unimaginable.

AI-assisted drones and robots that use computer vision are transforming civil engineering by boosting efficiency, safety, and precision. They quickly survey sites, create detailed maps and 3D models, ensure quality by scanning structural components, automate labor-intensive tasks, monitor safety hazards, analyze environmental factors, and create digital twins for better planning and project management. These technologies enable more informed decisions and streamline construction processes.

AI for Structural Analysis and Simulation

AI is transforming structural analysis by automating complex calculations and enhancing the accuracy of predictions for structural behavior under various loads. Machine learning algorithms learn from historical data, refining load calculations and structural assessments. AI-driven simulation tools like Ansys SimAI, used in the automobile industry, enable engineers to quickly test and optimize car designs by exploring numerous design alternatives, reducing time and cost for physical prototypes, and integrating AI with traditional methods for reliable and efficient designs.

IBM leverages AI in civil engineering for structural health monitoring. Its AI technology analyzes sensor data from infrastructure (bridges, buildings, tunnels) to detect and predict potential issues, ensuring safety and longevity.

For instance, IBM's Watson IoT platform combines AI with IoT sensors to monitor infrastructure health by collecting data on stress, strain, temperature, and vibration. AI processes this data to identify patterns and anomalies that indicate structural weaknesses or potential failures, enabling proactive maintenance and repairs to reduce catastrophic failure risks and extend infrastructure lifespan.

AI in Construction Management

AI tools can also be used for project scheduling and resource management. AI tools in construction management optimize project scheduling by analyzing historical data and predicting potential delays. These tools can also manage resources more efficiently, ensuring that materials and labor are allocated where they are needed most, reducing waste and improving productivity. AI-driven predictive analytics help construction managers identify and mitigate risks before they become critical issues. For example, Doxel is a tool that uses AI and autonomous robots to monitor construction sites, providing real-time progress tracking and identifying potential issues early.

Another example of a tool that excels in these areas is Procore. Procore is a comprehensive construction management platform that leverages AI to optimize project scheduling and resource management.

It analyzes historical data to predict potential delays and suggests adjustments to keep projects on track. Additionally, Procore efficiently manages resources by ensuring materials and labor are allocated where they are needed most, reducing waste and improving productivity.

AI for Sustainable Design
AI is instrumental in developing sustainable designs by optimizing energy use and reducing environmental impact. AI algorithms can analyze building performance data to suggest improvements in energy efficiency, material selection, and overall sustainability.

Tools like Autodesk's Insight use AI to provide real-time feedback on energy performance, helping architects design buildings that consume less energy and produce fewer emissions. AI can also simulate various environmental scenarios to ensure that buildings are resilient and adaptable to changing conditions.

AI in Urban Planning and Smart Cities

Credit: Sakina Syed

AI-driven tools are modernizing urban planning by analyzing vast amounts of data to optimize city layouts, transportation systems, and infrastructure. These tools can predict population growth, traffic patterns, and environmental impacts, helping planners create more efficient and livable cities.

In smart cities, AI is being utilized in various innovative ways. For instance, Barcelona employs AI for smart lighting and noise sensors to improve urban living conditions. The NEOM project in Saudi Arabia is another example, aiming to develop a fully AI-integrated city with optimized energy use, transportation, and infrastructure management. Additionally, Google's Green Light AI tool optimizes traffic light timings to reduce vehicle emissions and enhance traffic flow, and it is currently implemented in cities such as Bangalore, India, and Hamburg, Germany.

Future Trends in AI for Engineering and Architecture
In the Iron Man movies, Tony Stark engages in funny futuristic dialogues with his home assistant, an AI named Jarvis. While creating his Iron Man suits, Tony converses with Jarvis, who generates diagrams and designs based on Tony's verbal instructions. Although this appears to be utter science fiction, it is precisely the direction engineering researchers aspire to take the field in the future.

Emerging AI technologies, such as generative design, digital twins, AI-driven robotics, and VR and AR in design, are poised to revolutionize engineering and architecture. Virtual reality (VR) and augmented reality (AR) technologies, powered by AI, will provide immersive experiences that enhance architectural design and visualization. VR allows architects and clients to explore virtual models of buildings while AR overlays digital information onto the physical environment.

To prepare for this future and stay ahead, professionals in engineering and architecture must embrace continuous learning and upskilling. Mastering and utilizing AI tools will be essential for maintaining competitiveness and fostering innovation in the industry. For example, architects can take courses on AI-driven design software like Autodesk's Revit while engineers might explore AI applications in structural analysis through platforms like MATLAB.

Conclusion

AI is unlocking unprecedented potential in architecture and engineering, transforming visionary concepts into awe-inspiring structures. In engineering, AI applications range from predictive maintenance and quality control to optimizing complex systems and processes. In architecture, AI aids in generating innovative designs, optimizing space utilization and building costs, and ensuring sustainability.

Integrating AI into design and construction processes offers numerous benefits, including reduced costs, faster project completion, and improved safety. Imagine a future where every building is a testament to the seamless collaboration between human creativity and AI precision. Professionals can focus on more strategic and creative aspects of their work, ultimately leading to smarter, more sustainable, and aesthetically pleasing structures. In essence, AI is not just a tool but a partner in the creative and engineering processes, enabling professionals to achieve more than ever before.

With AI, the boundaries of what we can create are expanding, promising a world where our built environment is more innovative and inspiring than ever. As we continue to explore and harness the potential of AI, the future of engineering and architecture looks brighter, smarter, and more innovative. However, note that AI in these fields is already here, and it is advised to start implementing it today to stay ahead. The synergy between human ingenuity and AI technology is set to revolutionize the way we design and build, paving the way for a new era of intelligent, sustainable, and beautiful structures.

About the Author

Sakina is a dynamic senior data and AI consultant, AI engineer, and enthusiast with a stellar track record in the computer software industry, including notable tenures at tech giants like Microsoft. As a Microsoft AI engineer and Azure-certified professional, she excels in AI deployment, sales, management, teamwork, and leadership.

With a Bachelor of Science in Neuroscience and Mental Health Studies from the University of Toronto, Sakina has also enriched her expertise with courses in business management and IT. Her passion for writing and traveling adds a unique dimension to her professional persona.

Sakina's enthusiasm for business and the application of business psychology to understand consumer needs is matched by her extensive seven-year experience in customer service. She is adept at interpreting consumer feedback and statistical data, enabling her to identify market requirements with precision. Her strong passion for team management and leadership drives her dedication to delivering exceptional customer experiences and fostering business success.

Follow Sakina here:
LinkedIn: https://www.linkedin.com/in/sakina-syed/
Website: youraiconsultant.ca
Email: info@youraiconsultant.ca

CHAPTER 32

BREAKING THE LEGAL CODE: AI STRATEGIES FROM A TOP LEGAL TECHNOLOGIST

By Susan O. Tejuosho
Founder of Courtney Sessions
New York Metropolitan Area

> *I would argue that in the long term, failure to embrace and utilize AI may result in a lawyer not being the best and most competent advocate for their client.*
> —Michael Semanchik, Attorney at the California Innocence Project, 2023

The legal world is changing, and artificial intelligence (AI) is at the forefront of this transformation. As someone who has spent nearly a decade in legal services technology and AI-driven product development, I've witnessed firsthand how AI revolutionizes how we approach legal services. The legal profession has long been known for its adherence to tradition and precedent. This reverence for established practices, while crucial for maintaining the integrity of the legal system, has often re-sulted in a slower adoption of technological innovations compared to other industries. However, the landscape is changing rapidly, driven by advancements in AI and accelerated by global events like the COVID-19 pandemic. This chapter explores the journey of legal services from resis-tance to innovation to the current AI-driven revolution.

Historically, the legal profession has been cautious about adopting new technologies. This resistance stems from several factors such as:

- *Ethical Concerns*: the legal profession is bound by strict ethical rules due to confidentiality and philosophical considerations, which have made the legal profession and court systems cau-tious about adopting new technologies.

- *Risk Aversion*: the high stakes in legal matters have made prac-titioners hesitant to adopt nascent technologies. Lawyers are trained to minimize risk, which can lead to hesitancy.

- *Billable Hour Model*: the traditional billing structure in law firms, which requires attorneys to track every work done hourly for payment, has yet to incentivize efficiency improvements.

- *Regulatory Constraints*: legal practice is heavily regulated, which can slow innovation. Regulations governing law practice have also limited the adoption of specific technologies.

Despite these barriers, the legal industry has gradually embraced certain technologies, such as electronic filing systems for discovery and legal research databases. However, adoption has been slower than in other professional services industries, even with strict regulations like finance and healthcare.

Pandemic Acceleration and the Generative AI Breakthrough

The COVID-19 pandemic catalyzed digital transformation across indus-tries, and the legal sector was no exception. Seemingly overnight, courts, law firms, and legal departments were forced to adopt remote work tech-nologies and digital processes. This sudden shift demonstrated that the legal profession could adapt to technological changes when necessary. Key developments during this period included:

- *Virtual Courts*: many jurisdictions rapidly implemented virtual court proceedings, proving that justice could be administered effectively in a digital environment.

- *E-Signatures and Digital Documents*: the widespread adoption of electronic signatures and digital document management systems streamlined many legal processes.

- *Online Dispute Resolution*: platforms for online mediation and arbitration saw significant growth, paving the way for more accessible dispute resolution mechanisms.

During this phase, generative AI made significant progress, demonstrated by the release of ChatGPT in late 2022. While not specifically designed for legal applications, ChatGPT sparked interest by passing the bar exam designed for lawyers and increasing investment in GenAI for legal services.

This large language model demonstrated an unprecedented ability to understand and generate human-like text, including legal content. The impact of generative AI in law includes:

- *Legal Research*: AI models can now analyze vast amounts of legal documents, case law, and statutes in seconds, dramatically reducing research time.

- *Document Drafting*: AI can assist in drafting contracts, pleadings, and other legal documents, improving efficiency and consistency.

- *Predictive Analytics*: AI models can analyze past case outcomes to predict the likely results of current cases, aiding in strategic decision-making.

- *Client Interaction*: AI-powered chatbots can handle initial client inquiries, improving response times and freeing up lawyers for more complex tasks.

Making the Case for Access to Justice through AI

One of the most compelling arguments for adopting AI in legal services is its potential to improve access to justice. Due to high costs and complexity, the traditional legal system has often been criticized for needing to be more accessible to many. AI-powered tools have the potential to bridge this gap in several ways:

Self-Help Tools: AI-powered platforms can guide individuals through basic legal processes, such as filing simple court documents or understanding their legal rights, accelerating efficiency for self-represented litigants.

- *Document Automation*: AI can help create and complete legal documents, making the process more affordable and accessible.

- *Case Outcome Prediction*: AI tools can analyze past case data to predict potential outcomes, helping individuals make informed decisions about pursuing legal action.

- *Language Translation*: AI can break down language barriers in legal proceedings, ensuring equal access and legal simplification for non-native speakers.

- *Triage Systems*: AI can help legal aid organizations and courts efficiently triage cases, ensuring limited resources are allocated to those most in need.

Responsible AI Considerations

The potential role of AI-powered legal assistants in increasing access to justice, particularly for underserved communities and individuals who may face barriers to traditional legal services, carries the potential to revolutionize legal services virtually. However, as with any transformative technology, they also present significant challenges that must be carefully navigated to ensure responsible and ethical development. Some key considerations include:

Transparency and Accountability: AI chatbots in legal services must be designed with transparency, ensuring users understand they are interacting with an AI system and providing clear explanations of the chatbot's capabilities and limitations.

- *Privacy and Data Protection*: legal AI chatbots must adhere to strict privacy and data protection protocols, safeguarding user information and ensuring confidentiality, especially in sensitive conflict resolution contexts.

- *Bias Mitigation*: legal AI chatbots should be trained on diverse datasets and undergo rigorous testing to mitigate potential biases that could undermine their ability to provide fair and impartial guidance.

- *Human Oversight and Control*: while AI chatbots can assist in legal information and management, human oversight and control are essential, particularly in complex or high-stakes situations where human judgment and decision-making are critical.

- *Ethical Principles and Guidelines*: the development and deployment of AI legal assistants should adhere to established ethical principles and guidelines, ensuring responsible and beneficial use of the technology.

AI Enterprise Examples in Legal Services

The legal industry is seeing rapid adoption of AI tools across various domains, with prominent companies such as:

Harvey: this is an AI-powered legal research tool backed by OpenAI, which law firms use to streamline research processes and draft documents.

- *Thomson Reuters CoCounsel*: this AI legal assistant can perform legal research, document review, and contract analysis.

- *GCAI:* An AI copilot for in-house lawyers, that asks questions, through documents or links to work with general counsels to create briefs and company policies.

- *JusticeText*: this AI tool helps public defenders analyze body camera footage and other audiovisual evidence.

By leveraging artificial intelligence (AI) to bridge the gap, Justice-Text's legal technology showcases how AI improves access to justice. It equips public defenders with the tools needed to function as effectively as attorneys with the vast resources of large law firms. AI-powered tools like JusticeText and legal assistants help mitigate significant disadvantages in the criminal justice system caused by errors stemming from the overwhelming volume of documents in trials. This is especially crucial for small-scale legal practices, nonprofit legal organizations, and rural courts with limited funding, which may need more resources to hire additional paralegals or prosecutors to review evidence, meet with witnesses, or obtain expert knowledge. Without such resources, the risk of mistrials or wrongful convictions increases.

Leading law firms are not just adopting AI tools; many top firms actively invest in innovation. For example:

- *Reed Smith*: this firm launched an innovation hub to develop new legal service delivery models. Their innovation hub, Reed Smith Global Solutions, focuses on legal technology and innovation, including AI applications.

- *Cleary Gottlieb*: The firm has established a legal technology group to explore and implement AI solutions in their practice.

- *A&O Sherman*: this firm developed its suite of AI tools to provide technology, resourcing, and end-to-end solutions for its clients and to deliver legal services in new and flexible ways.

- *Dentons*: this firm created Nextlaw Labs, which acts as a global collaborative innovation platform for investing in and developing new technologies for the legal profession. These initiatives demonstrate a shift in the legal industry's attitude toward embracing innovation and technology as a core part of legal practice.

In response to the growing use of AI in legal proceedings, alternative court services are equally innovating. The American Arbitration Association has AAAiLab. This center harnesses the benefits of artificial intelligence through guidance, education, and products, as well as a new AI clause for arbitration agreements. This clause addresses the use of AI tools in arbitration proceedings, ensuring transparency and fairness.

The Growth of AI Solutions Roles in Legal Services

The rise of AI in legal services has led to the creation of some new roles within law firms and legal departments, such as:

- *Legal Engineers*: these are professionals who bridge the gap between law and technology, designing and implementing AI solutions.

- *Legal Data Scientists*: these experts apply data science and machine learning techniques to legal data, deriving insights and developing predictive models.

- *Legal Innovation Officers*: these leaders drive the adoption and implementation of new technologies, including AI, within law firms and legal departments.

- *AI Ethics Officers*: these specialists ensure the ethical use of AI in legal contexts, addressing issues of bias, transparency, and accountability.

- These roles reflect the growing recognition of AI as a crucial component of modern legal practice and the increasing importance of technology skills in the legal profession.

AI Governance in Legal Services

As AI becomes more prevalent in legal services, it's crucial to ensure its responsible use with:

- *Bias Mitigation*: ensuring AI systems don't perpetuate or exacerbate existing biases in the legal system.

- *Transparency*: making the decision-making processes of AI systems in legal applications as transparent and explainable as possible to maintain trust in the legal system.

- *Data Privacy*: protecting client information's confidentiality and sensitive legal data and complying with data protection regulations, such as the National Institute of Standards and Technology (NIST) AI risk management framework for AI systems and the EU AI Act.

- *Ethical Guidelines*: Developing and adhering to ethical guidelines for AI use in legal practice. The American Bar Association (ABA) issued formal guidance for use of generative AI, which has provided new guidelines for AI (1). Initiatives like Responsible AI in Legal Services (RAILS) by Duke Center for Law and Tech, which I am privileged to be one of their consultants, serve the growing need to create a space and opportunity for legal to move forward with AI while keeping the integrity of legal services (2).

In legal services, there's a predominant need for a system design involving humans-in-the-loop and not toward fully autonomous AI agents that function without human intervention, given the nature of authorized practice and unauthorized practice of law.

AI researchers have developed a framework for analyzing the integration of human experiences with data by utilizing human-in-the-loop methods to measure the level of meaningful human involvement required for necessary chatbot guidelines and output training (3). Stanford's human-centered artificial intelligence states how such automation flaws can be managed: "It's clear there is something worth preserving in many of the things we do in life, which is why automation can't be reduced to a simple binary between 'manual' and 'automatic'" (4). Instead, it's about searching for the right balance between aspects that we find helpful in automating versus tasks in which it remains meaningful for us to participate. Human feedback allows AI to produce a more grounded and centered intelligence, where humans approximate how machines function, not machines.

Challenges and Future Outlook

Despite the progress, challenges remain in the widespread adoption of AI in legal services. Skills gap is a significant challenge, with many legal professionals needing more technical skills to leverage AI tools fully (5). When I started considering my legal technologist career, I took a design thinking class at Columbia University School of Professional Studies, where I learned about designing prototypes and testing new product development. Over time, as I grew, I went to Y Combinator Startup School to learn about building products from the leading tech experts. I took the Replit 100 Days of Code, learning programming languages like Python and HTML and eventually got a certification with Google Cloud for Advanced Generative AI Developers and Responsible AI.

More universities are actively trying to upskill law students and lawyers. For example, Oxford University offers AI foundation classes tailored to lawyers for a firm grasp of artificial intelligence's technical, legal, and ethical aspects. Harvard University also provides online executive education in computer science courses for lawyers. Topics include programming languages, algorithms, cybersecurity, cloud computing, database design, and challenges at the intersection of law and technology (6). Stanford Law School created a center for legal informatics called CodeX, emphasizing the research and development of computational law (complaw). This branch of legal informatics is concerned with the mechanization of legal reasoning. They aim to provide researchers, lawyers, entrepreneurs, and technologists with work side by side to advance the frontier of legal technology, bringing new levels of legal efficiency, transparency, and access to legal systems worldwide (7).

The Future of AI in Legal Services

As we look to the future, several trends will likely shape AI's continued evolution in legal services, including:

Increased Automation: The rise of AI agents will make routine legal tasks will become increasingly automated, allowing legal professionals to focus on higher-value work.

- *Personalized Legal Services*: AI will enable personalized legal services tailored to individual client needs and preferences.

- *Predictive Justice*: advanced AI models may be more significant in predicting case outcomes and informing legal strategies.

- *Regulatory Technology (RegTech)*: AI will increasingly be used to ensure compliance with complex and evolving regulations.

- *Transformation of Legal Education*: law schools will increasingly incorporate AI and technology into their curricula.

- *Blockchain and Smart Contracts*: integrating AI with blockchain technology may revolutionize contract execution and enforcement.

The legal services industry is poised for a technological revolution driven by AI. While the journey from resistance to innovation has been extended, the pace of change is accelerating rapidly. As we embrace these new technologies, balancing innovation with ethical considerations and the fundamental principles of justice is crucial. The future of legal services lies in the thoughtful integration of AI, enhancing human expertise rather than replacing it and ultimately making justice more accessible to all.

The transformation of legal services through AI represents a paradigm shift in law practice, and the rise of alternative legal service providers (ALSPs) exemplifies the transition from traditional legal practice to legal tech. From its historical resistance to change, the legal profession has embarked on a journey of innovation accelerated by global events and technological breakthroughs. As AI continues to evolve, it promises to make legal services more efficient, accessible, and equitable.

However, this transformation is challenging. Ensuring AI's responsible and ethical use in legal contexts will be crucial. The legal profession must navigate this new landscape carefully, balancing the benefits of innovation with the fundamental principles of justice and fairness that underpin the legal system. As we move forward, the success of AI in legal services will depend not just on technological advancements but on the ability of legal professionals to adapt, innovate, and maintain their crucial role as guardians of justice in an increasingly digital world.

References

(1) https://www.americanbar.org/news/abanews/aba-news-archives/2024/07/aba-issues-first-ethics-guidance-ai-tools/

(2) https://rails.legal/

(3) I. Lage, A. Ross, S. J. Gershman, B. Kim, and F. Doshi-Velez, "Human-in-the-loop interpretability prior," Adv. Neural Inf. Process. Syst., vol. 31, 2018. [Online]. Available: https://research.google/pubs/human-in-the-loop-interpretability-prior/

(4) https://hai.stanford.edu/news/humans-loop-design-interactive-ai-systems

(5) https://www.law.com/2023/03/20/lawyers-must-become-ai-literate-now/

(6) https://pll.harvard.edu/course/computer-science-lawyers

(7) https://law.stanford.edu/codex-the-stanford-center-for-legal-informatics/

(8) https://solve.mit.edu/challenges/unbundle-policing-accelerator/solutions/54085

About the Author

Susan Tejuosho is an accomplished lawyer with a unique blend of legal expertise, technological prowess, and a deep passion for leveraging AI to drive innovation in the legal industry. With a strong foundation in law, holding an LLB and an LLM in international law, Susan is a 2024 nominee at the American Legal Technology Award for Access to Justice. She is also at the forefront of leading infrastructure and scaling technology development for responsible AI in legal services with Duke's Center for Law & Tech. Driven by the vision of making conflict resolution more accessible, she got into MITSolve Incubator and founded Courtney Sessions to design AI-powered legal tech solutions, including a platform that streamlines onboarding processes and a groundbreaking AI chatbot framework for conflict resolution. This journey to legal tech has given her unique insights into the challenges and opportunities of innovating in the legal sector.

Email: susantejuosho@outlook.com
LinkedIn: https://www.linkedin.com/in/susantejuosho/

CHAPTER 33

REDEFINING VITALITY: THE FUTURE OF HEALTH, LONGEVITY, AND AI-DRIVEN INNOVATION

By Martin Wyss, PhD
Tech & Health Entrepreneur, AI Expert
Zurich, Switzerland

> *The best way to predict the future of health is to create it.*
> —Abraham Lincoln (adapted for healthcare)

The Dawn of AI-Enhanced Healthcare

Artificial intelligence is rapidly changing medical practice. Many studies have outlined the impact of AI on practically all levels, from diagnostics to improving treatment outcomes. Places like Stanford Medicine, Memorial Sloan Kettering, and Mayo Clinic have been front-runners in showing the real benefits of AI in clinical usage. Stanford Medicine's study published in *Nature Medicine* outlines the role of AI in improving diagnostic accuracy and efficiency.

The above studies indicated that AI could diagnose pneumonia with as high as 92% accuracy through chest X-rays, reducing diagnostic time by 30% in radiology departments (Conley, 2023). This is a typical example of how AI supports better decision-making and early detection of life-critical conditions for improved patient outcomes. In preventive cardiology, the Apple Heart Study, in collaboration with Stanford, enrolled 419,297 participants and showed that AI-powered wearable devices can correctly detect atrial fibrillation, resulting in timely medical intervention with an impressive 84% positive predictive value (Perez, 2020).

Memorial Sloan Kettering Cancer Center has applied AI to strengthen cancer detection and treatment planning in oncology. In an article in *Nature Medicine*, for example, they showed that the analysis of pathology images by AI matched or outperformed human experts in diagnostic accuracy, improving cancer diagnosis and treatment outcomes significantly (Taylor et, 2024). Moreover, Mayo has systematically embedded AI into its clinical practices to enhance diagnostic precision, speed treatment, and make better use of the resources available, thus enhancing patient outcomes (Halamka, 2021). Many other countries have sought the National Health Service and the World Health Organization's use of AI to leverage AI's power to deliver better healthcare services at a national level. The NHS has deployed AI to decrease waiting times for diagnostic procedures, manage screening programs more effectively, and enhance resource allocation (Barclay & Smith, 2023). Similarly, reports from the WHO indicate that AI enhances diagnostic capabilities and improves access to expertise, thus decreasing disparities in healthcare.

While improvements have been achieved, challenges still stand in integrating AI relating to data, systems integration, and training (World Health Organization, 2021). Though the National Academy of Medicine has identified these challenges, their ongoing research and development continue to overcome them. While looking to the future, new AI technologies in advanced imaging, precision medicine, and predictive analytics promise to further transform healthcare delivery, enhance patient outcomes, and minimize costs (National Academies of Sciences, 2021).

The Transition from Reactive to Preventive Healthcare

The enlightenment of shifting people from a reactive to a proactive healthcare system is comprehensively legible, as well as the supporting research and concept demonstrations. Many studies have shown how the transition enhances health benefits and health costs. Preventive healthcare aims to find and address the risk before diseases are fully manifested or even escalated with the help of fluencies, including AI and data analytics (Google Health, n.d.). One of the best examples of preventive healthcare solutions is a diabetic retinopathy detection AI system created by Google Health. Analyses of clinical trials conducted by the same authors in The Lancet Digital Health showed that this system yielded 90% sensitivity and 98% specificity for detecting vision-threatening diabetic retinopathy (Google Health, n.d.). Whenever used in real-life environments in both India and Thailand contexts, the system was able to identify this disease prior to its onset manifested in symptoms, thus resulting in early treatment and better outcomes for patients. Similarly, Kaiser Permanente's study in the *New England Journal of Medicine* details how AI is effectively applied as a predictive method to reduce hospital readmission. This system identifies high-risk patients and improves chronic disease management, reducing readmissions by 25% and enhancing overall care (Greene, 2020).

More importantly, a comparative study on environmental health by Mount Sinai published in *Nature Medicine* epitomizes better the role of care in prevention. Analyzing records from more than 5,000 participants, this research outlined the patterns of environmental exposures that lead to the risk of disease and, thus, pressed the alarm for focused preventive interventions (Mount, 2023). The long-standing Framingham Heart Study, embedding AI analysis, has helped identify many features critical for cardiovascular disease prevention, including very early warning signs and interventions that could prevent heart disease (Hong, 2018). Similarly, the CDC has made equal strides in preventive care through predictive analytics by empowering outbreak detection and identifying at-risk populations for better resource optimization (CDC, 2024). The Precision Prevention Initiative by the National Institutes of Health brought personalized prevention with genetic risks and targeted interventions. Indeed, this would ensure better health outcomes, exemplifying success in personalized care.

While the use of AI in medicine is highly promising, many challenges exist, including challenges around system adaptation, provider training, and access to preventive healthcare (National Institutes of Health, 2020). The National Academy of Medicine has identified and documented various successful strategies to overcome those challenges. New technologies such as advanced biosensors and predictive analytics will further facilitate this future of preventive healthcare in terms of better efficacy and coverage within the regulatory oversight provided by the FDA (Raaga Likhitha Musunuri & Bhatt, 2024).

Personalized Health Optimization through AI

Artificial intelligence in personalized healthcare is no longer seen merely as support; it is also extensively supported by peer-reviewed studies and real-world implementations. The plethora of innovations includes the continuous glucose monitoring systems developed by Dexcom and Abbott that have completely changed the dimensions of managing diabetes (Miller, 2020). According to research findings appearing in *Diabetes Care*, these systems result in improvement in glycemic control, reduce HbA1c by an average of 0.5%, decrease time spent in hypoglycemia, and improve the overall quality of life of individuals with type 1 and type 2 diabetes (Miller, 2020). More specifically, studies at Stanford University published in *Nature Medicine* demonstrate how machine learning algorithms can predict a person's glycemic response to different foods. This indicates that identical foods can provoke different glucose responses among subjects, hence the need for personalized nutrition.

The Mayo Clinic has also capitalized on AI-driven personalization, where research in the *Journal of the American Medical Association* details how using AI analytics on patient data leads to more accurate dosing of medication, better predictions of treatment responses, and improved patient outcomes (O'Hara, 2019). Similarly, large-scale analyses, such as the UK Biobank's review of 500,000 participants' data, have provided critical insights into genetic changes influencing individual responses to exercise, nutrition, medication, and sleep. Thus, individual responses and the best health advice are related to a genetic link (UK Biobank, n.d.).

In sports, research by the International Olympic Committee for the *Journal of Sports Sciences* demonstrates that injury rates, recovery, and athletic performance are improved through personalized nutrition after customized AI-driven training programs (Luo et al., 2018).

Harvard Medical School's Precision Medicine Initiative and Cleveland Clinic's Precision Medicine Program further highlight the importance of artificial intelligence in tailoring the selection of treatments, predicting drug responses, and optimizing outcomes in chronic diseases and cancer (Daneshjou et al., n.d.). The All of Us Research Program at the National Institutes of Health brings together over one million participants, providing data on how individual variation influences disease susceptibility and treatment responses. This fosters further development in personalized health. Although data privacy, algorithm validation, and equity of access remain, the emergent research, FDA regulations, and even the economic benefits of the World Health Organization help develop personalized health to ensure better outcomes and reduce costs (Warraich et al., 2024).

Evidence-Based AI Mental Health Applications

AI applications for mental health have shown considerable promise in both clinical and real-world settings. One such clinically validated AI application, Woebot, an AI chatbot, has been proven to reduce symptoms of depression effectively. A 2017 study published in *JMIR Mental Health* found that college students using Woebot reported a significant decrease in depression symptoms over two weeks compared with a control group with high levels of the therapeutic bond (Fitzpatrick et al., 2017). The NIMH has discussed the growing role of AI in the mental health screening and monitoring process, primarily via natural language processing through detecting the linguistic markers of depression and anxiety from texts at accuracy rates comparable to that of methods involving human assessments.

Clinically, the REACH VET program at the Veterans Health Administration applies machine learning to medical records in order to identify veterans with a high risk of suicide. Studies have demonstrated a 20% increase in mental healthcare utilization by high-risk veterans (Whitepaper, 2019). At Mount Sinai, using machine learning algorithms to triage mental health patients led to improved efficiency in patient assessment, with wait times reduced by 30%, as reported in *Nature Digital Medicine* (Whitepaper, 2019). The IAPT program in the NHS in England has also applied AI-powered assessment tools to increase access to therapy services, serving more than 25% of its population (NHS Digital, 2022).

They even cover things like emotional distress identification, such as Google's machine learning research on voice. These AI tools have been designed with algorithms that can identify the underlying emotional state with near-perfect precision of a clinician. The American Psychiatric Association also raised concerns about privacy, clear guidelines used by professionals, and the producer's relationship with the patient in using AI in mental healthcare. Internationally, AI mental health applications are used for screening purposes in 34 countries and crisis intervention in 28 countries (American Psychiatric Association, 2018). According to the World Health Organization's 2021 Digital Health Report, services have been rendered less costly and more accessible via these tools. Other issues involve access and outcome verification, which are a few of the challenges future studies are tackling, and integrating the use of AI in treating mental health conditions allows the possibility of a mix of technology and clinical professionals (World Health Organization, 2023).

Privacy, Data Security, and Ethical Considerations

Integrating AI in healthcare has raised severe privacy, data security, and ethics issues. As more personal health data is gathered and analyzed by the healthcare systems, the need for protection measures has become highly imperative. According to a 2023 study by the Pew Research Center, 79% of Americans are not confident that companies will take responsibility for misusing personal health data (Alder, 2024). This concern is further enforced by the alarming number of healthcare data breaches, over 3,700 incidents from 2009 to 2020, affecting 268 million patient records (Alder, 2024). Organizations such as the Mayo Clinic have, thus, moved to implement differential privacy techniques to overcome these challenges. In this method, noise is introduced into the data, hindering the identification of individuals while maintaining statistical value for research into the data (Halamka, 2021). The EU's General Data Protection Regulation has imposed global standards on health data protection, including explicit consent, the right to be forgotten, and strict breach notification protocols. The UK's NHS has met this standard through its Data Security and Protection Toolkit (NHS Digital, 2022).

If implemented, federated learning can provide a way to study AI models developed by Google Health to detect breast cancer. It allows training across several institutions without necessarily sharing the raw patient data (Google Health, n.d.). It better preserves privacy for patients while improving model generalizability. According to a report from the World Health Organization in 2021, there are unequal distributions between high-income and low-income countries (World Health Organization, 2021). The FDA has designed regulatory frameworks involving premarket review, monitoring of performance, and periodic updates for healthcare applications using AI. Even the NIST has established cybersecurity frameworks regarding health data on access control, encryption, and incident response (Warraich et al., 2024).

The AI-driven Veterans Health Administration system represents ethical issues regarding AI in healthcare for predicting suicide risk. That is promising but raises concerns about intervention timing and resource allocation. Organizations like IEEE developed ethical guidelines about transparency, accountability, and privacy protection. In the future, international collaboration will be necessary as AI integrates into healthcare in order for it to continue being ethical, secure, and fair for all (Chatila et al., 2017).

Integrating AI with Human Intuition and Wisdom

AI in healthcare does not replace human intuition and wisdom but enhances them. Several studies have demonstrated that AI systems improve diagnostic efficiency and accuracy when combined with human expertise. In a study published in *Nature Medicine*, radiologists' performance improved when given AI support; they outperformed either artificial intelligence or human clinicians working in isolation, with an increase of 11.5% in the accuracy of mammogram screenings and a reduction of 5.7% in false positives (Siwicki, 2021). Stanford Medicine AI tools save physicians' time from performing routine tasks and, thus, enable them to have more face-to-face interactions with patients, boosting patient satisfaction and enabling better care.

The Mayo Clinic has also shown many benefits that can be derived from the combination of AI analytic capability with expert medical judgment through the use of AI within its clinical workflows. Its AI-assisted platform supports clinicians in analyzing patient data, flagging potential medication interactions, and identifying candidates for clinical trials (Scherting, 2023). By reducing the administrative analysis burden by about 32%, the system offers superior disease detection rates early (Scherting, 2023).

Cleveland Clinic has put in place a systematic approach to incorporating AI while ensuring that the doctor-to-patient relationship remains intact; however, expanding to intensify data and pattern identification is critical.

AI has many advantages, but difficulties still exist in resistance to change, extensive training, and system integrations. These challenges have been addressed through phased implementations and standardized workflows, significantly increasing AI adoption (Scherting, 2023). As AI in health continues to evolve, the focus will be further developed on personalization, improvement of predictive capability, and enhanced support in clinical decision-making without losing the irreplaceable human touch of care.

The Future of AI in Public Health and Community Well-Being

AI is increasingly at the heart of designing a future for public health and community well-being. Indeed, cities such as Singapore and Copenhagen provide potent examples of how AI-driven innovation can shape urban health systems. For example, Singapore's Smart Nation initiative deploys hundreds of environmental sensors across the city-state to monitor, in real time, air quality, urban heat patterns, and disease indicators, significantly strengthening the city-state's potential for detection and intervention in health crises. Throughout the COVID-19 pandemic, the TraceTogether system has been an excellent example of how AI could mobilize tremendous public health responses while considering citizens' privacy (The Singapore TraceTogether Story for COVID-19 Contact Tracing, n.d.). Copenhagen tapped into AI for pedestrian flow pattern analysis and air quality data, leading to an uptick of 42% in active transportation and a reduction of 27% in hotspots of air pollution (Bettman, 2018).

AI-driven workplace systems, such as Microsoft's intelligent building technologies, have successfully maintained environmental conditions to improve employee well-being with quantifiable health and cognitive benefits. Population health management programs powered by AI, such as those by NHS Digital, to identify at-risk groups and improve vaccination rates create better health outcomes among under-resourced populations.

Notwithstanding these developments, the so-called digital divide remains challenging because many low-income communities lack the infrastructure to embrace AI-driven health programs and policies (NHS

Digital, 2022). The World Health Organization calls for equity in access to AI technologies, adding that ethical frameworks will ensure just implementation. Integrating AI with public health strategy will continue to hold promise for healthier, more resilient communities while posing challenges in equity and access.

References

Alder, S. (2024, January 18). *December 2023 Healthcare Data Breach Report. HIPAA Journal.* https://www.hipaajournal.com/december-2023-healthcare-data-breach-report/

American Psychiatric Association. (2018). *Clinical Practice Guidelines* | psychiatry.org. Psychiatry.org. https://www.psychiatry.org/psychiatrists/practice/clinical-practice-guidelines

Barclay, S., & Smith, C. (2023, June 23). *£21 million to roll out artificial intelligence across the NHS.* GOV.UK. https://www.gov.uk/government/news/21-million-to-roll-out-artificial-intelligence-across-the-nhs

Bettman, K. T. (2018, October 9). *A breath of fresh air: Measuring air quality in Copenhagen.* Google. https://blog.google/around-the-globe/google-europe/breath-fresh-air-measuring-air-quality-copenhagen/

CDC. (2024, May 8). *Insight Net.* Insight Net. https://www.cdc.gov/insight-net/php/about/index.html

Chatila, R., Firth-Butterflied, K., Havens, J. C., & Karachalios, K. (2017). The IEEE Global Initiative for Ethical Considerations in Artificial Intelligence and Autonomous Systems [Standards]. *IEEE Robotics & Automation Magazine*, 24(1), 110–110. https://doi.org/10.1109/mra.2017.2670225

Conley, M. (2023, November 10). *How Stanford Medicine is capturing the AI moment.* Stanford Medicine Magazine. https://stanmed.stanford.edu/translating-ai-concepts-into-innovations/

Daneshjou, R., Brenner, S., Chen, J., Crawford, D., Finlayson, S., Kidziński, Ł., & Bulyk, M. (n.d.). *Precision Medicine: Using Artificial Intelligence to Improve Diagnostics and Healthcare.* Retrieved November 21, 2024, from https://psb.stanford.edu/psb-online/proceedings/psb22/intro_pm.pdf

Fang, J., Chen, W., Hou, P., Liu, Z., Zuo, M., Liu, S., Feng, C., Han, Y., Li, P., Shi, Y., & Shao, C. (2023). NAD+ metabolism-based immunoregulation and therapeutic potential. *Cell & Bioscience*, *13*(1). https://doi.org/10.1186/s13578-023-01031-5

Fitzpatrick, K. K., Darcy, A., & Vierhile, M. (2017). Delivering Cognitive Behavior Therapy to Young Adults With Symptoms of Depression and Anxiety Using a Fully Automated Conversational Agent (Woebot): A Randomized Controlled Trial. *JMIR Mental Health*, *4*(2). https://doi.org/10.2196/mental.7785

Google Health. (n.d.). *Google Health*. Health.google. https://health.google/health-research/imaging-and-diagnostics/

Greene, J. (2020, November 11). *Real-time in-hospital alerts associated with lower patient mortality*. Kaiser Permanente Division of Research. https://divisionofresearch.kaiserpermanente.org/blog/2020/11/11/real-time-in-hospital-alerts/

Halamka, J. (2021, January 22). *How is AI Impacting Health Care Today?—Mayo Clinic Platform*. Mayo Clinic Platform. https://www.mayoclinicplatform.org/2021/01/22/how-is-ai-impacting-health-care-today/

Hong, Y. (2018, April 16). *Framingham Heart Study (FHS) | National Heart, Lung, and Blood Institute (NHLBI)*. Nih.gov. https://www.nhlbi.nih.gov/science/framingham-heart-study-fhs

Johns Hopkins Medicine. (2019, February 8). *Johns Hopkins Medicine, based in Baltimore, Maryland*. Hopkinsmedicine.org. https://www.hopkinsmedicine.org/

Luo, J., Meng, Q., & Cai, Y. (2018). Analysis of the Impact of Artificial Intelligence Application on the Development of Accounting Industry. *Open Journal of Business and Management*, *06*(04), 850–856. https://doi.org/10.4236/ojbm.2018.64063

McElvery, R. (2020, September 17). *Bringing new energy to mitochondria research*. MIT Department of Biology. https://biology.mit.edu/bringing-new-energy-to-mitochondria-research/

Miller, E. M. (2020). Using Continuous Glucose Monitoring in Clinical Practice. *Clinical Diabetes*, *38*(5), 429–438. https://doi.org/10.2337/cd20-0043

Mount. (2023). *Mount Sinai Institute for Exposomics Research Awarded $8.45 Million Grant to Study Environmental Health.* Mount Sinai Health System. https://www.mountsinai.org/about/newsroom/2023/mount-sinai-institute-for-exposomics-research-awarded-eight-million-grant-to-study-environmental-health

National Academies of Sciences, E. (2021). Implementing High-Quality Primary Care: Rebuilding the Foundation of Health Care. In nap.nationalacademies.org. https://nap.nationalacademies.org/catalog/25983/implementing-high-quality-primary-care-rebuilding-the-foundation-of-health

National Institutes of Health. (2020, February 5). *The Promise of Precision Medicine.* National Institutes of Health (NIH). https://www.nih.gov/about-nih/what-we-do/nih-turning-discovery-into-health/promise-precision-medicine

NHS Digital. (2022, November 14). *General Data Protection Regulation (GDPR) - Information.* NHS Digital. https://digital.nhs.uk/data-and-information/keeping-data-safe-and-benefitting-the-public/gdpr

NHS Digital. (2022, September 29). *Psychological Therapies, Annual report on the use of IAPT services, 2021-22.* NHS Digital. https://digital.nhs.uk/data-and-information/publications/statistical/psychological-therapies-annual-reports-on-the-use-of-iapt-services/annual-report-2021-22

O'Hara, J. (2019, November 7). *How artificial intelligence and machine learning is changing medicine.* Mayo Clinic News Network. https://newsnetwork.mayoclinic.org/discussion/how-artificial-intelligence-and-machine-learning-is-changing-medicine/

Perez, M. (2020, February 26). *Apple Heart Study: Not just for atrial fibrillation.* Medical Conferences. https://conferences.medicompublishers.com/specialization/cardiology/2020-02-27-144515/

Raaga Likhitha Musunuri, & Bhatt, A. (2024). Navigating the Future of Healthcare. *Advances in Healthcare Information Systems and Administration Book Series,* 73–106. https://doi.org/10.4018/979-8-3693-8552-4.ch003

Scherting, K. (2023, April 17). *How Mayo Clinic Is Using Artificial Intelligence to Transform Cardiovascular Care - Mayo Clinic*

Magazine. Mayo Clinic Magazine. https://mayomagazine.mayoclinic.org/2023/04/mayo-clinic-using-artificial-intelligence-to-transform-cardiovascular-care/

The Singapore TraceTogether Story for COVID-19 Contact Tracing. (n.d.). SMU Newsroom. https://news.smu.edu.sg/news/2022/04/08/singapore-tracetogether-story-covid-19-contact-tracing

Siwicki, B. (2021, May 10). *Mass General Brigham and the future of AI in radiology*. Healthcare IT News. https://www.healthcareitnews.com/news/mass-general-brigham-and-future-ai-radiology

Taylor, M. (2024, June 4). *Memorial Sloan Kettering, NCI, creates AI that predicts cancer treatment outcomes*. Beckershospitalreview.com. https://www.beckershospitalreview.com/oncology/memorial-sloan-kettering-nci-create-ai-that-predicts-cancer-treatment-outcomes.html

UK Biobank. (n.d.). *About our data*. Www.ukbiobank.ac.uk. https://www.ukbiobank.ac.uk/enable-your-research/about-our-data

Walton, M. (2023, December 18). *The Science of Aging and Longevity*. Massachusetts General Hospital Giving; Massachusetts General Hospital. https://giving.massgeneral.org/stories/the-science-of-aging-and-longevity

Warraich, H. J., Tazbaz, T., & Califf, R. M. (2024). FDA Perspective on the Regulation of Artificial Intelligence in Health Care and Biomedicine. *JAMA*. https://doi.org/10.1001/jama.2024.21451

Whitepaper, C. (2019). *OFFICE OF INFORMATION AND TECHNOLOGY*. https://digital.va.gov/wp-content/uploads/2023/01/CIO WhitepaperArtificialIntelligence.pdf

World Health Organization. (2021, June 28). *WHO issues first global report on Artificial Intelligence (AI) in health and six guiding principles for its design and use*. Www.who.int. https://www.who.int/news/item/28-06-2021-who-issues-first-global-report-on-ai-in-health-and-six-guiding-principles-for-its-design-and-use

World Health Organization. (2023, October 19). *WHO outlines considerations for regulation of artificial intelligence for health*. Www.who.int. https://www.who.int/news/item/19-10-2023-who-outlines-considerations-for-regulation-of-artificial-intelligence-for-health

About the Author

Dr. Martin Wyss, founder and CEO of Clinic X, is a visionary entrepreneur redefining healthcare through the integration of integrative medicine, advanced bioinformatics, and artificial intelligence. Driven by a lifelong curiosity about health and human potential, Wyss has dedicated his career to uncovering root causes of disease and crafting personalized solutions for optimal well-being.

With a doctorate in integral medicine, an executive MBA, and current PhD research in bioinformatics and engineering, Wyss combines deep scientific knowledge with cutting-edge technology. His expertise spans integrative medicine, environmental health, and precision diagnostics, including groundbreaking research on exosomes and brain health.

Wyss's global collaborations and personal health journey have shaped Clinic X's transformative approach, offering solutions that address physical, mental, and environmental factors. Through innovation and compassion, Wyss envisions a future where healthcare empowers individuals not only to overcome illness but to thrive at their highest potential.

Email: m.wyss@clinicx.ch
LinkedIn: https://www.linkedin.com/in/martin-wyss/

APPENDIX A

LIST OF POPULAR AI TOOLS

If you're looking for an AI tool that you can use for various tasks, below are a selection of popular ones. (Note: The creation of this list was assisted by ChatGPT)

ChatGPT (OpenAI)
Use case: Conversational AI, content generation, coding assistance, customer support.
Description: A large language model that can perform tasks like answering questions, generating text, and providing creative writing prompts.
URL: https://chat.openai.com

DALL·E (OpenAI)
Use case: Image generation from text prompts.
Description: An AI model that creates images based on text descriptions, useful in art, marketing, and design.
URL: https://openai.com/dall-e

Microsoft Copilot (Microsoft 365)
Use case: Productivity assistance in Microsoft apps.
Description: An AI tool integrated into Microsoft 365 applications like Word, Excel, PowerPoint, and Outlook to assist with tasks such as drafting, data analysis, and summarizing.
URL: https://www.microsoft.com/en-us/microsoft-365/copilot

Grok (X)
Use case: Conversational AI for Twitter.
Description: An AI chatbot integrated into Twitter (X), designed to assist users in real-time by answering questions, generating tweets, and providing insights, all in a conversational manner.
URL: https://x.com

Gemini (Google)
Use case: Conversational AI and language model.
Description: A multimodal AI model that handles text, images, and other input types. It is expected to enhance search, productivity, and various creative tasks with advanced language understanding.
URL: https://about.google/intl/en/products/gemini/

Claude (Anthropic)
Use case: Conversational AI and assistant tasks.
Description: A language model focused on safe and reliable conversational interactions, used for answering questions, providing insights, and assisting in creative or technical tasks.
URL: https://www.anthropic.com/product

Perplexity
Use case: AI-powered search engine and question-answering.
Description: An AI search engine designed to answer user queries with concise and sourced information, offering both traditional search results and direct answers to questions.
URL: https://www.perplexity.ai

Phind
Use case: AI-powered search engine for developers.
Description: A search engine designed to answer complex technical and programming questions by providing relevant, code-focused answers from documentation, forums, and other resources.
URL: https://www.phind.com

Otter.ai
Use case: Transcription and meeting notes.
Description: A tool that provides real-time transcription of meetings, interviews, and conversations, with AI-generated notes, summaries, and key points.
URL: https://otter.ai

Duolingo (AI-enhanced version)
Use case: Language learning.
Description: A language learning app that uses AI to personalize lessons, predict learner needs, and provide conversational practice with virtual characters.
URL: https://www.duolingo.com

MidJourney
Use case: Artistic and visual content creation.
Description: A tool that generates high-quality art and illustrations from text prompts, used heavily in creative industries.
URL: https://www.midjourney.com

Jasper AI
Use case: Copywriting and marketing content.
Description: An AI-driven platform that helps in generating marketing copy, blog posts, product descriptions, and social media content.
URL: https://www.jasper.ai

Synthesia
Use case: AI-generated video creation.
Description: A tool that lets users create videos with AI avatars and voiceovers, often used for training videos, marketing, and education.
URL: https://www.synthesia.io

Lumen5
Use case: Video content creation.
Description: This AI-powered tool transforms blog posts and articles into shareable videos by automatically matching text with visuals.
URL: https://www.lumen5.com

Grammarly
Use case: Writing assistance and grammar checking.
Description: An AI tool that provides real-time grammar, punctuation, and style suggestions to improve writing quality.
URL: https://www.grammarly.com

Notion AI
Use case: Productivity and organization.
Description: An extension to Notion, it helps with summarizing content, generating ideas, and automating writing processes within a workspace.
URL: https://www.notion.so/product/ai

Pictory
Use case: Video creation.
Description: An AI tool that converts long-form content into short, engaging videos using AI-generated scenes and text overlays.
URL: https://pictory.ai

Hugging Face
Use case: Machine learning model hosting and deployment.
Description: A platform for hosting, sharing, and deploying machine learning models, particularly useful in natural language processing and computer vision.
URL: https://huggingface.co

Runway ML
Use case: Creative AI for video editing, art, and design.
Description: A platform providing creative AI tools for video editing, animation, and design, widely used by artists and filmmakers.
URL: https://runwayml.com

Copy.ai
Use case: Content generation.
Description: An AI-driven platform that generates marketing copy, product descriptions, and other written content for businesses.
URL: https://www.copy.ai

GitHub Copilot (OpenAI)
Use case: Code generation and software development.
Description: An AI-powered code completion tool integrated into code editors like Visual Studio Code, helping developers write code faster by suggesting whole lines or blocks of code.
URL: https://github.com/features/copilot

APPENDIX B

GLOSSARY OF COMMON AI TERMS

Many of the terms below are beyond the scope of this book. For someone new to AI, just reading through these terms will greatly expand your vision of the workings of AI. This glossary can also provide a starting point for anyone wanting to look for further information about AI. (Note: The creation of this glossary was assisted by ChatGPT)

Activation Function
A mathematical function used in neural networks to determine the output of a node (neuron). Activation functions introduce non-linearity, allowing neural networks to solve more complex problems.

AI-Assisted Creativity
The use of AI tools to enhance human creativity. This includes generating art, music, writing, or design based on user input, making the creative process faster and more accessible.

AI Compass
A strategic framework that guides individuals or organizations in how to ethically and effectively use AI technologies. It helps align AI development with long-term goals, ethical standards, and societal values.

AI-Driven Insights
The ability of AI to analyze large datasets and extract meaningful patterns or trends. These insights help businesses and individuals make data-driven decisions faster and more accurately.

AI GamePlan
A structured plan or strategy that outlines how AI will be developed, integrated, and utilized in specific projects or across an organization. An AI GamePlan typically includes objectives, resources, timelines, and risk management.

AI Governance
The policies, frameworks, and structures that ensure the responsible development, deployment, and management of AI systems. AI governance includes legal, ethical, and operational standards to minimize risks and ensure fairness and transparency.

AI Integration
The process of embedding AI technologies into existing business processes, tools, or workflows. AI integration aims to improve efficiency, accuracy, and decision-making across various industries.

AI-Powered Automation
Using AI to automate repetitive or time-consuming tasks. This can range from chatbots responding to customer inquiries to AI systems automating complex business processes, saving time and resources.

AI Transparency
The practice of making AI systems and their decision-making processes understandable to users. Transparency helps build trust in AI by allowing users to see how and why a model arrived at a certain conclusion.

Algorithm
A set of rules or instructions that a computer follows to perform a task or solve a problem. In AI, algorithms are used to process data, learn from patterns, and make decisions.

Artificial General Intelligence (AGI)

A type of AI that can understand, learn, and apply knowledge across a wide range of tasks at a human-like level. Unlike narrow AI, which excels at specific tasks, AGI would have general cognitive abilities akin to human intelligence.

Augmented Intelligence

An approach where AI is designed to assist and enhance human decision-making rather than replace it. Augmented intelligence focuses on collaboration between humans and AI systems for better outcomes.

Autonomous AI Agents

AI-powered systems that can make independent decisions and take actions without human intervention. Think of AI personal assistants that can plan your schedule, book travel, or even manage entire workflows autonomously.

Backpropagation

An algorithm used to train neural networks by adjusting the weights of connections between neurons. Backpropagation helps minimize errors by propagating them backward through the network.

Bias in AI

Refers to the presence of prejudiced or unfair outcomes in AI models due to biased training data or flawed algorithms. Bias in AI can lead to discrimination or unfair treatment in areas like hiring, lending, and law enforcement.

Convolutional Neural Networks (CNNs)

A type of neural network particularly effective for image and video analysis. CNNs use convolutional layers to capture spatial hierarchies in data, making them well-suited for tasks like object detection and image classification.

Conversational AI

AI systems designed to engage in human-like dialogue. These systems are often used in chatbots, virtual assistants, and customer service applications to respond to user queries and hold natural conversations.

Deep Learning
A type of machine learning that uses neural networks with many layers (hence "deep"). Deep learning excels in processing large, complex datasets and is used in tasks like image and speech recognition, natural language processing, and more.

Edge AI
AI that processes data on local devices (e.g., smartphones, cameras) rather than in the cloud. Edge AI enables faster processing and reduces the need for constant internet connectivity, making it suitable for real-time applications.

Epoch
A term used in machine learning to describe one complete cycle through the entire training dataset. Multiple epochs are used during training to improve the model's accuracy.

Ethics in AI
The principles and guidelines that govern the responsible development and use of AI technologies. Ethical AI focuses on fairness, transparency, accountability, and ensuring AI does no harm to individuals or society.

Explainable AI (XAI)
Techniques and tools that allow humans to understand how AI models make decisions. XAI is important for increasing transparency and trust, particularly in critical applications like healthcare and finance.

Fine-Tuning
A process in machine learning where a pre-trained model is further trained on a smaller, task-specific dataset. Fine-tuning helps adapt a general-purpose model to a specialized application.

Generative AI
AI models designed to generate new content such as text, images, audio, or video. These models can create outputs that resemble human-made content by learning from vast datasets.

Gradient Descent

An optimization algorithm used to minimize errors in machine learning models by adjusting model parameters (like weights in neural networks). It is essential for training models by finding the best possible performance on a task.

Human-in-the-Loop (HITL)

A process where human judgment is integrated with AI decision-making to improve performance. HITL ensures that human oversight guides AI systems in tasks requiring nuanced understanding or critical outcomes.

Hyperparameters

Settings or configurations that define the structure and learning process of a machine learning model (e.g., learning rate, number of layers). Hyperparameters are tuned to optimize model performance.

Inference

The process of applying a trained AI model to new data to make predictions or decisions. Inference is what happens when an AI model is used in real-world applications after training.

Large Language Models (LLMs)

AI models, like GPT-4, that process and generate human-like text. They are trained on massive datasets of text to understand and predict language patterns, enabling a wide range of applications like answering questions, translation, and content generation.

Loss Function

A mathematical function that quantifies the difference between a model's predictions and the actual outcomes. The goal of training is to minimize the loss function, improving the model's accuracy.

Machine Learning (ML)

A subset of AI where algorithms improve their performance at tasks by learning from data, without being explicitly programmed for each specific task. It involves training models to recognize patterns and make decisions based on data.

Natural Language Processing (NLP)
A field of AI focused on enabling computers to understand, interpret, and generate human language. It encompasses tasks like translation, sentiment analysis, summarization, and conversation generation.

Natural Language Understanding (NLU)
A subfield of NLP focused on enabling machines to comprehend the meaning and intent behind human language. NLU is critical for tasks like question answering, sentiment analysis, and dialogue systems.

Neural Networks
Computational models inspired by the human brain, composed of layers of interconnected nodes ("neurons"). Neural networks are the foundation of most AI models and are used to recognize patterns, classify data, and make decisions.

Neural Radiance Fields (NeRFs)
A groundbreaking approach to generating 3D models from simple 2D images, used in virtual reality, digital twins, and high-quality 3D rendering.

Nirvana State
A theoretical concept in AI where a system reaches a level of perfection in its performance, free of bias, errors, and inefficiencies. Achieving this state would mean that the AI operates at optimal levels in all aspects, but it is largely seen as an unattainable ideal.

Overfitting
A situation where a machine learning model learns the training data too well, including its noise and errors. Overfitting causes poor performance on new, unseen data, as the model fails to generalize.

Personalization
The use of AI to tailor experiences to individual users, often by analyzing user data and preferences. Personalization is common in recommendation systems (e.g., Netflix, Spotify) and digital marketing.

Prioritization Matrix

A tool used in decision-making to rank AI projects or tasks based on factors such as urgency, impact, and resources. This matrix helps teams focus on high-priority initiatives and allocate resources effectively.

Prompt Engineering

The practice of designing and refining inputs (prompts) for AI models, particularly large language models, to elicit the desired output. It's key to making generative AI models perform specific tasks effectively.

Quantum Machine Learning

The integration of AI with quantum computing has the potential to solve complex problems far beyond the capabilities of classical computing, with breakthroughs in materials science and cryptography.

Recurrent Neural Networks (RNNs)

A neural network model designed to process sequential data, such as time series or natural language. RNNs have a memory component that helps them maintain context over sequences, often used in speech recognition and language modeling.

Reinforcement Learning

An area of machine learning where an agent learns to make decisions by interacting with its environment and receiving rewards or penalties based on its actions. This method is often used in gaming, robotics, and complex decision-making scenarios.

Reinforcement Learning from Human Feedback (RLHF)

A type of reinforcement learning where AI models are trained and fine-tuned using human input on their outputs. Humans provide feedback, which the model uses to adjust its actions and improve over time, particularly for tasks requiring nuanced judgment.

Supervised Learning

A machine learning approach where a model is trained on labeled data. The model learns to map inputs to specific outputs by observing examples (e.g., classifying images of cats and dogs).

Training Data
The dataset used to teach an AI model. During training, the model learns from this data to make predictions or decisions. The quality and quantity of training data significantly impact the model's performance.

Transfer Learning
A machine learning technique where a model trained on one task is adapted to perform a different but related task. This allows for faster training and improved performance by leveraging previously learned knowledge.

Transformers
A neural network architecture designed to handle sequential data but more efficiently than RNNs. Transformers are the foundation of many modern AI models, including large language models, by enabling parallel processing of data sequences.

Underfitting
Occurs when a machine learning model is too simple to capture the underlying patterns in the data. An underfitted model performs poorly on both the training data and new data.

Unsupervised Learning
A machine learning technique where models are trained on unlabeled data. The goal is to find hidden patterns or groupings in the data without explicit instructions (e.g., clustering similar items).

DID YOU ENJOY THIS BOOK?

If you enjoyed reading this book, you can help by suggesting it to someone else you think might like it, and **please leave a positive review** wherever you purchased it. This does a lot in helping others find the book. We thank you in advance for taking a few moments to do this.

THANK YOU

You might also like other Thin Leaf Press titles:

The AI Mindset: Thriving Within Civilization's Next Big Disruption

Peak Performance: Mindset Tools for Managers

Peak Performance: Mindset Tools for Sales

Peak Performance: Mindset Tools for Leaders

Peak Performance: Mindset Tools for Business

Peak Performance: Mindset Tools for Entrepreneurs

Peak Performance: Mindset Tools for Athletes

The Successful Mind: Tools to Living a Purposeful, Productive, and Happy Life

The Successful Body: Using Fitness, Nutrition, and Mindset to Live Better

The Successful Spirit: Top Performers Share Secrets to a Winning Mindset

Winning Mindset: Elite Strategies for Peak Performance

Winner's Mindset: Peak Performance Strategies for Success

The Life Coach's Tool Kit, Vol. 1

The Life Coach's Tool Kit, Vol. 2

The Life Coach's Tool Kit, Vol. 3

Ordinary to Extraordinary

The Magical Lightness of Being

Explore.